Indoor Exposure to Fine Particulate Matter and Practical Mitigation Approaches

PROCEEDINGS OF A WORKSHOP

David A. Butler and Joe Alper, Rapporteurs

National Academy of Engineering

Board on Population Health and Public Health Practice
Health and Medicine Division

The National Academies of
SCIENCES · ENGINEERING · MEDICINE

THE NATIONAL ACADEMIES PRESS
Washington, DC
www.nap.edu

THE NATIONAL ACADEMIES PRESS   500 Fifth Street, NW   Washington, DC 20001

This activity was supported by Contract No. 68HERC19D0011, Order No. 68HERC20F0334 between the National Academies of Sciences, Engineering, and Medicine and the US Environmental Protection Agency (EPA). Any opinions, findings, conclusions, or recommendations expressed in this publication do not necessarily reflect the views of any organization or agency that provided support for the project. Views expressed in written conference materials or publications and by speakers and moderators do not necessarily reflect the official policies of the EPA, nor does any mention of trade names, commercial practices, or organization imply endorsement by the United States Government.

International Standard Book Number-13: 978-0-309-26328-3
International Standard Book Number-10: 0-309-26328-X
Digital Object Identifier: https://doi.org/10.17226/26331

Additional copies of this proceedings are available from the National Academies Press, 500 Fifth Street NW, Keck 360, Washington, DC 20001; (800) 624-6242 or (202) 334-3313; www.nap.edu.

Copyright 2022 by the National Academy of Sciences. All rights reserved.

Printed in the United States of America

Suggested citation: National Academies of Sciences, Engineering, and Medicine. 2022. *Indoor Exposure to Fine Particulate Matter and Practical Mitigation Approaches: Proceedings of a Workshop*. Washington, DC: The National Academies Press. https://doi.org/10.17226/26331.

The National Academies of
SCIENCES · ENGINEERING · MEDICINE

The **National Academy of Sciences** was established in 1863 by an Act of Congress, signed by President Lincoln, as a private, nongovernmental institution to advise the nation on issues related to science and technology. Members are elected by their peers for outstanding contributions to research. Dr. Marcia McNutt is president.

The **National Academy of Engineering** was established in 1964 under the charter of the National Academy of Sciences to bring the practices of engineering to advising the nation. Members are elected by their peers for extraordinary contributions to engineering. Dr. John L. Anderson is president.

The **National Academy of Medicine** (formerly the Institute of Medicine) was established in 1970 under the charter of the National Academy of Sciences to advise the nation on medical and health issues. Members are elected by their peers for distinguished contributions to medicine and health. Dr. Victor J. Dzau is president.

The three Academies work together as the **National Academies of Sciences, Engineering, and Medicine** to provide independent, objective analysis and advice to the nation and conduct other activities to solve complex problems and inform public policy decisions. The Academies also encourage education and research, recognize outstanding contributions to knowledge, and increase public understanding in matters of science, engineering, and medicine.

Learn more about the National Academies of Sciences, Engineering, and Medicine at **www.nationalacademies.org**.

The National Academies of
SCIENCES · ENGINEERING · MEDICINE

Consensus Study Reports published by the National Academies of Sciences, Engineering, and Medicine document the evidence-based consensus on the study's statement of task by an authoring committee of experts. Reports typically include findings, conclusions, and recommendations based on information gathered by the committee and the committee's deliberations. Each report has been subjected to a rigorous and independent peer-review process and it represents the position of the National Academies on the statement of task.

Proceedings published by the National Academies of Sciences, Engineering, and Medicine chronicle the presentations and discussions at a workshop, symposium, or other event convened by the National Academies. The statements and opinions contained in proceedings are those of the participants and are not endorsed by other participants, the planning committee, or the National Academies.

For information about other products and activities of the National Academies, please visit www.nationalacademies.org/about/whatwedo.

PLANNING COMMITTEE ON THE INDOOR EXPOSURE TO FINE PARTICULATE MATTER AND PRACTICAL MITIGATION STRATEGIES WORKSHOP[1]

RICHARD L. CORSI (*Chair*), Dean of Engineering, University of California, Davis
SEEMA BHANGAR, Senior Indoor Air Quality Manager, WeWork
WANYU R. CHAN, Research Scientist and Deputy Indoor Environment Group Leader, Energy Analysis and Environmental Impact Division, Lawrence Berkeley National Laboratory
ELIZABETH C. MATSUI, Professor of Population Health and Pediatrics and Associate Director of the Health Transformation Research Institute, Dell Medical School at the University of Texas at Austin
LINDA A. MCCAULEY, Professor and Dean of Emory University's Nell Hodgson Woodruff School of Nursing
KIMBERLY A. PRATHER, Professor in Chemistry and Biochemistry, Scripps Institution of Oceanography, University of California, San Diego
DAVID Y. PUI, Regents Professor and the L.M. Fingerson/TSI Inc. Chair in Mechanical Engineering and Director, Particle Technology Laboratory and Center for Filtration Research, University of Minnesota, Minneapolis
JEFFREY A. SIEGEL, Professor of Civil Engineering and Member, Building Engineering Research Group, University of Toronto
MARINA E. VANCE, Assistant Professor and McLagan Family Faculty Fellow, Department of Mechanical Engineering, University of Colorado Boulder

National Academies Staff

DAVID A. BUTLER, J. Herbert Hollomon Scholar
GURU MADHAVAN, Norman R. Augustine Senior Scholar and Senior Director of Programs
KATHLEEN STRATTON, Scholar
COURTNEY HILL, Associate Program Officer
CAMERON H. FLETCHER, Editor

---

[1] National Academies of Sciences, Engineering, and Medicine planning committees are solely responsible for organizing the workshop, identifying topics, and choosing speakers. The responsibility for the published proceedings rests with the workshop rapporteurs and the institution.

MICHAEL HOLZER, Senior Program Assistant (through June 2021)
MAIYA SPELL, Senior Program Assistant (from June 2021)

Consultant

JOE ALPER, Consulting Writer

Acknowledgments

This Proceedings of a Workshop was reviewed in draft form by individuals chosen for their diverse perspectives and technical expertise. The purpose of this independent review is to provide candid and critical comments that will assist the National Academies of Sciences, Engineering, and Medicine in making each published proceedings as sound as possible and to ensure that it meets the institutional standards for quality, objectivity, evidence, and responsiveness to the charge. The review comments and draft manuscript remain confidential to protect the integrity of the process.

We thank the following individuals for their review of this proceedings:

Diane Gold, Harvard University
Andrew Persily, National Institute of Standards and Technology
Mark Utell, University of Rochester

Although the reviewers listed above provided many constructive comments and suggestions, they did not see the final draft of the proceedings before its release. The review of this proceedings was overseen by Michael Ladisch, Purdue University. He was responsible for making certain that an independent examination of this proceedings was carried out in accordance with institutional procedures and that all review comments were carefully considered. Responsibility for the final content of this proceedings rests entirely with the rapporteurs and the National Academies.

Contents

ACRONYMS AND ABBREVIATIONS   xvii

1 INTRODUCTION   1
Conduct of the Workshop, 3
Organization of the Proceedings, 4

2 OUTDOOR SOURCES OF INDOOR PARTICULATE MATTER   5
Indoor Particulate Matter of Outdoor Origin and the Disparities in Sources and Exposures across Communities, 5
Outdoor-to-Indoor Transport Mechanisms and Particle Penetration for Fine Particulate Matter, 11
Outdoor Particulate Matter Sources and the Chemical Transformations That Take Place When They Interact with the Indoor Environment, 16
Discussion, 20

3 INDOOR SOURCES OF INDOOR PARTICULATE MATTER   25
Fine Particulate Matter Emissions from Cooking, 25
Secondary Aerosol Formation of Fine Particulate Matter in the Indoor Environment, 32
The Effect of Humidity on the Chemistry and Biology of Indoor Air, 35

The Influence of Sources of Indoor Fine Particulate Matter
  on the Characterization of Exposure and Evaluation
  of Health Effects, 37
Discussion, 40

4   DAY ONE SUMMARY   43

5   HEALTH EFFECTS OF EXPOSURE TO INDOOR
    PARTICULATE MATTER   45
    The Overall (Mostly Cardiovascular) Health Burden
      of Indoor $PM_{2.5}$ Exposure, 46
    Pulmonary Disease Associated with Fine Particular Matter
      Exposure in Indoor Environments and Disparities
      in Economically Challenged Communities, 51
    Wildfire Smoke and Other Ambient Air Pollution Comes
      Indoors: Health Effects and the Building Characteristics
      That Mitigate Them, 55
    Discussion, 58

6   INDOOR EXPOSURE TO PARTICULATE MATTER:
    METRICS AND ASSESSMENT   61
    Transcending Complexity: Indoor $PM_{2.5}$ Measurement,
      Exposure, and Control, 61
    The Challenge of Moving from the Measurement of Indoor
      $PM_{2.5}$ to Evaluating Occupant Exposure, 68
    The Utility, Use, and Misuse of Low-Cost Consumer
      Indoor Particulate Matter Sensors, 72
    Discussion, 75

7   DAY TWO SUMMARY   79

8   INDOOR PARTICULATE MATTER EXPOSURE
    CONTROL AND MITIGATION   81
    $PM_{2.5}$ Filtration and Air Cleaning in Residential
      Environments, 81
    $PM_{2.5}$ Exposure Control in Schools, 87
    Mitigation of $PM_{2.5}$ Exposure Associated with Cooking, 91
    Discussion, 94

9	**OCCUPANT RESPONSES TO INDOOR PARTICULATE MATTER**	99
	Portable Indoor Air Cleaners and Human Behavior, 99	
	How Building Occupants Interpret and Respond to Indoor Air Quality Sensor Data, 105	
	Public Health Responses to Reduce Community Exposure to Indoor $PM_{2.5}$, 108	
	Discussion, 112	
10	**WORKSHOP SUMMARY AND CLOSING REFLECTIONS**	117
	Outdoor $PM_{2.5}$, 118	
	Health Effects of Indoor $PM_{2.5}$, 119	
	Mitigation of Indoor $PM_{2.5}$, 120

REFERENCES 123

APPENDIXES
A Workshop Agenda 139
B Biographic Sketches of Planning Committee Members and Workshop Speakers 147

Figures

2-1 The diverse origins of ambient $PM_{2.5}$, 6
2-2 Top five sources of $PM_{2.5}$ emissions in the San Bernardino and Muscoy, California, community, actual and projected, 9
2-3 3-year (2013–15) average of the 24-hour $PM_{2.5}$ concentrations for the United States network of outdoor $PM_{2.5}$ monitoring stations, 12
2-4 Average $PM_{2.5}$ concentrations on an annual basis, 2000–12, estimated by combining modeling and remote sensing information, 13
2-5 Infiltration of outdoor particulate matter into the indoor environment, 14
2-6 Output from a NASA model showing the different sources of particulate matter, 17

3-1 High particle concentrations observed during cooking activities, with and without ventilation, 27
3-2 Total indoor particulate matter mass concentrations during the HOMEChem Thanksgiving Day experiment, 28
3-3 Indoor residential activities enhance semivolatile organic compound concentrations, 30
3-4 Particulate matter mass deposited in the respiratory system in different contexts, 31

3-5	Change in solute concentrations at varying relative humidities, 37	
3-6	Framework for determining contributors to indoor particulate matter exposure disparities, 39	
5-1	Global burden of disease attributable to 20 leading risk factors in 2010, 47	
5-2	Overview of diseases, conditions, and biomarkers affected by outdoor air pollution, 48	
5-3	Possible mechanistic paths linking particulate matter exposure and cardiovascular disease, 49	
5-4	Indoor particulate pollution and asthma morbidity, 52	
5-5	Solid fuel use as primary heating source, 54	
5-6	US regional estimates for weeks with a risk of very large fires in the mid-21st century compared to the end of the 20th century, 56	
6-1	One person's $PM_{2.5}$ exposure in Singapore on June 25, 2013, during a wildfire smoke event, 62	
6-2	Size and spatial complexity of particle deposition in different regions of the lung, 63	
6-3	Temporal complexity's correlation with occupancy, using the example of optical particle counter data from a university classroom in normal use, 64	
6-4	Measurements of particulate matter in a controlled chamber over time, 65	
6-5	EPA environmental health paradigm, 68	
6-6	Variability of $PM_{2.5}$ exposures by person, day, and microenvironment, 70	
8-1	Home air volumes that pass through a filter when the heating, ventilation, and air conditioning system is operating in residences sampled in 3 studies in selected North American locations, 82	
8-2	Fraction of time that a heating, ventilation, and air conditioning system operates, 83	
8-3	Single-pass filtration removal of particles at different face velocities, 84	
8-4	Results of in situ testing of four types of air filters compared to laboratory results, 85	
8-5	Percentage of students attending a school within 250 meters of a major roadway (2005–06 school year), 88	

9-1 Conceptual diagram showing elements of behavior change to reduce particulate matter exposure and improve health, 101
9-2 Seasonal summary of particulate matter concentrations for filter use study control and intervention groups, 103
9-3 Occupant use of an air cleaner (filter) during and between monitoring periods, 104
9-4 Occupant perception of indoor environmental quality, including air quality, 106

Acronyms and Abbreviations

$\mu g/m^3$	microgram per cubic meter
ASHRAE	American Society of Heating, Refrigerating and Air-Conditioning Engineers
COPD	chronic obstructive pulmonary disease
EPA	US Environmental Protection Agency
HEPA	high-efficiency particulate air
HOMEChem	House Observations of Microbial and Environmental Chemistry
HVAC	heating, ventilation, and air conditioning
MERV	minimum efficiency reporting value (a measure of a filter's ability to capture particles between 0.3 and 10 microns)
NASEM	National Academies of Sciences, Engineering, and Medicine
PCAP	persistent cold air pooling
$PM_{0.1}$	ultrafine aerosols

$PM_{2.5}$	fine particulate matter
PM_{10}	coarse particulate matter
SVOC	semivolatile organic compound
VOC	volatile organic compound

1

Introduction[1]

Overwhelming evidence exists that exposure to outdoor fine particulate matter ($PM_{2.5}$)[2] is associated with a range of short-term and chronic health impacts, including asthma exacerbation, acute and chronic bronchitis, heart attacks, increased susceptibility to respiratory infections, and premature death (Chen et al., 2007), with the burden of these health effects falling more heavily on underserved and marginalized communities (Mikati et al., 2018; Parker et al., 2018). Although less studied to date, indoor exposure to $PM_{2.5}$ is also gaining attention as a potential source of adverse health effects, particularly given that Americans spend 90 percent of their lives indoors (Klepeis et al., 2001) and indoor $PM_{2.5}$ levels can exceed outdoor levels (Logue et al., 2011). $PM_{2.5}$ found indoors can be particles of outdoor origin that migrate indoors or originate from indoor sources such as cooking, candle burning, and other occupant activities, including cleaning.

To better understand the sources of indoor $PM_{2.5}$, the possible health effects of exposure to indoor $PM_{2.5}$, and engineering approaches and interventions to reduce those exposure risks, the National Academies of Sciences, Engineering, and Medicine (National Academies) convened a virtual workshop with the following statement of task:

---

[1] This proceedings has been prepared by the workshop rapporteurs as a factual summary of what occurred at the workshop. Statements, recommendations, and opinions expressed are those of individual presenters and participants and are not necessarily endorsed or verified by the National Academies of Sciences, Engineering, and Medicine, and they should not be construed as reflecting any group consensus.

[2] "Fine particulate matter" and "$PM_{2.5}$" are used interchangeably throughout this document.

The National Academies shall convene a planning committee of scientific experts to conduct a workshop on the state of the science on exposure to fine particulate matter ($PM_{2.5}$) indoors, its health impacts, and engineering approaches and interventions to reduce exposure risks, including practical mitigation solutions in residential settings. The workshop will feature invited presentations and panel discussions on these topics. It shall include consideration of
1. the key implications of scientific research and engineering practice for public health, including potential near-term opportunities for incorporating what is known into practice; and
2. where additional research will be most critical to understanding indoor exposure to $PM_{2.5}$ and the effectiveness of interventions.

Opportunities for advancing research by addressing methodological and technological barriers and enhancing coordination and collaboration between the science, medical, and engineering communities will also be given attention. The indoor environments considered in this workshop will be limited to nonindustrial exposure within buildings.

The resulting workshop, "Indoor Exposure to Fine Particulate Matter and Practical Mitigation Approaches," was held on April 14, 21, and 28, 2021 and was sponsored by the US Environmental Protection Agency's (EPA) Office of Radiation and Indoor Air. It was intended in part to further the discussion about indoor $PM_{2.5}$ that began at the 2016 National Academies workshop, *Health Risks of Indoor Exposure to Particulate Matter: Workshop Summary* (NASEM, 2016).[3] As planning committee chair Richard Corsi (University of California, Davis) noted, the committee decided to focus the scope of these workshops on exposures that occur in residential and school buildings and on existing and practical mitigation technologies and approaches rather than novel, emerging, or unproven technologies. In keeping with its statement of task, the planning committee also strove to center attention on $PM_{2.5}$, although the workshop did address some sources of both $PM_{2.5}$ and larger particles.[4]

In his introduction to the workshop, Jonathan Edwards (director, EPA Office of Radiation and Indoor Air) explained that EPA's indoor air program was launched and authorized by statute in the 1980s in response to the growing concern in the scientific and public health communities about the risks associated with poor indoor air quality. Since then, EPA has developed a robust, nonregulatory program that works to reduce exposure to high-priority, indoor contaminants, including radon, indoor asthma

---

[3] A summary of the 2016 workshop and additional resources are available at https://www.nap.edu/catalog/23531/health-risks-of-indoor-exposure-to-particulate-matter-workshop-summary.

[4] $PM_{2.5}$ refers to particles equal to or smaller than 2.5 microns in diameter. Ultrafine particles, which the speakers also discussed, are a subset of $PM_{2.5}$, just as $PM_{2.5}$ is a subset of PM_{10}, or particles 10 microns in diameter and smaller.

triggers, volatile organic compounds, environmental tobacco smoke, biological contaminants, and particulate matter, in homes, schools, and other commercial buildings.

Edwards noted that the COVID-19 pandemic has increased the public's awareness of the importance of indoor air quality. He added that since the 2016 workshop—which highlighted the fact that indoor exposure to $PM_{2.5}$ is a serious public health issue—EPA, the research community, and indoor environment-related industries have made substantial progress in developing tools and resources that can help reduce the risks posed by exposure. Examples of these tools and resources include revised EPA guidance on residential air cleaners and filters (US EPA, 2021), improved filter and air cleaning technologies, additional analyses of the health effects of indoor-generated $PM_{2.5}$, and the increasing use of sensor-based technologies to detect $PM_{2.5}$.

Concluding his remarks, Edwards said EPA will use the results of the workshop to continue to build a scientific foundation on the impacts of indoor exposure to particulate matter. The workshop will also help EPA communicate the importance of the science, to develop and share new and revised guidance on engineering and behavioral approaches, to mitigate exposure to indoor particulate matter, and to help drive advances in indoor particulate matter mitigation technology.

CONDUCT OF THE WORKSHOP

The three-webinar workshop (see Appendix A for the agenda) was organized by an independent planning committee in accordance with National Academies procedures.[5] The planning committee members were Seema Bhangar, Wanyu R. Chan, Richard L. Corsi (chair), Elizabeth C. Matsui, Linda A. McCauley, Kimberly A. Prather, David Y. Pui, Jeffrey A. Siegel, and Marina E. Vance (see Appendix B for biographic sketches of the committee members and workshop speakers). The workshop was broadcast live online. Webcast analytics reported more than 600 unique views for each of the three broadcasts, with observers in nearly all the US states and over 30 countries on 6 continents. The workshop presentations were subsequently posted to the Web along with links to the videos of the talks.[6]

---

[5] The role of the planning committee was limited to planning the workshop.

[6] Available for **April 14:** https://www.nationalacademies.org/event/04-14-2021/indoor-exposure-to-fine-particulate-matter-and-practical-mitigation-approaches-workshop-on-sources-of-indoor-fine-particulate-matter, **April 21:** https://www.nationalacademies.org/event/04-21-2021/indoor-exposure-to-fine-particulate-matter-and-practical-mitigation-approaches-workshop-on-indoor-pm-exposure-health-metrics-and-assessment, and **April 28:** https://www.nationalacademies.org/event/04-28-2021/indoor-exposure-to-fine-particulate-matter-and-practical-mitigation-approaches-workshop-on-mitigation-of-indoor-exposure-to-fine-particulate-matter.

ORGANIZATION OF THE PROCEEDINGS

This publication summaries the presentations and discussions that took place during the workshop; it is divided into nine chapters plus supporting appendices. Chapters 2 and 3 describe the major outdoor and indoor sources of $PM_{2.5}$, respectively, and Chapter 4 recaps the first day's presentations and discussion. Chapter 5 discusses the health effects of exposure to indoor $PM_{2.5}$, Chapter 6 explores the measurement of $PM_{2.5}$ and the challenges of evaluating an individual's actual exposure to $PM_{2.5}$, and Chapter 7 summarizes the second day's presentations and discussions. Chapter 8 examines engineering approaches to control and mitigate indoor $PM_{2.5}$, and Chapter 9 reviews how individuals respond to information about their possible exposure to indoor $PM_{2.5}$ and public health approaches to reduce community exposure to indoor $PM_{2.5}$. Chapter 10 summarizes the workshop's key points as noted by the workshop planning committee chair.

Several speakers offered personal observations regarding actions that might be taken by individuals or governmental entities. However, in accordance with National Academies policies, the planning committee did not attempt to establish any conclusions or recommendations about needs and future research directions, focusing instead on issues identified by individual speakers and workshop participants. This proceedings was drafted by rapporteur Joe Alper in collaboration with National Academies staff member David A. Butler as a factual summary of what occurred at the workshop, and the National Academies does not endorse or verify the statements.

The National Academies is, at EPA's behest, conducting a study[7] that will address the issues discussed in the workshop in greater detail and will offer findings, conclusions, and recommendations regarding them. This study will be released in early 2023.

---

[7] Health Risks of Indoor Exposure to Fine Particulate Matter and Practical Mitigation Solutions: https://www.nationalacademies.org/our-work/health-risks-of-indoor-exposures-to-fine-particulate-matter-and-practical-mitigation-solutions.

2

Outdoor Sources of Indoor Particulate Matter

In the workshop's first session, three panelists described some of the major outdoor sources of indoor fine particulate matter ($PM_{2.5}$). Cesunica Ivey (University of California, Riverside) discussed what some of those sources are and disparities in exposure to those sources across communities. Brent Stephens (Illinois Institute of Technology) reviewed the mechanisms by which outdoor $PM_{2.5}$ gets indoors, and Delphine Farmer (Colorado State University) then talked about the chemical transformations that occur when outdoor $PM_{2.5}$ interacts with the indoor environment. An open discussion moderated by Kimberly Prather (Scripps Institution of Oceanography at the University of California San Diego) followed the three presentations.

INDOOR PARTICULATE MATTER OF OUTDOOR ORIGIN AND DISPARITIES IN SOURCES AND EXPOSURES ACROSS COMMUNITIES

$PM_{2.5}$ arises from multiple sources including transportation, industrial, and agricultural activities; wildfire smoke, and (in coastal locations) marine vessels (Figure 2-1), and models are an important tool for understanding the contributions of those various sources to ambient $PM_{2.5}$, said Cesunica Ivey. As a graduate student at the Georgia Institute of Technology, Ivey developed a hybrid source apportionment model that can attribute primary and secondary particulate matter in ambient air to different sources (Ivey et al., 2015). Her model combines a chemical transport model, a nonlinear optimization method, and spatial and temporal interpolations of adjustment

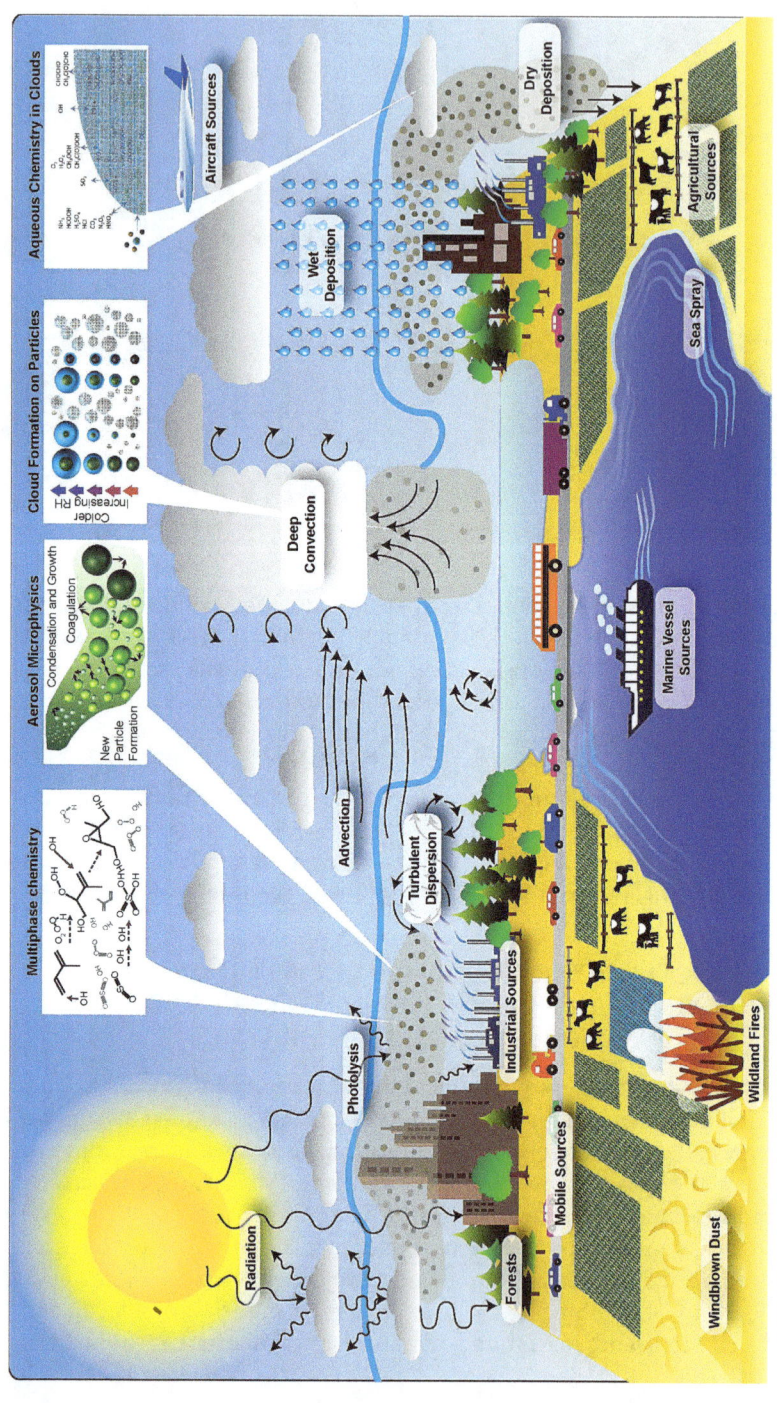

FIGURE 2-1 The diverse origins of ambient PM$_{2.5}$.
SOURCE: Ivey Slide 4 (original source: EPA, https://www.epa.gov/cmaq/overview-science-processes-cmaq).

factors to improve the estimates of source impacts across several different sources.

As Ivey explained, methods such as these can be useful for attributing secondary $PM_{2.5}$[1] in the model back to ground sources during meteorological phenomena such as persistent cold air pooling (PCAP) events, which occur frequently in the Western United States during winter and specifically in mountain towns. In the Salt Lake Valley, for example, a hybrid apportionment model found that natural gas and gasoline-powered mobile vehicles were the biggest contributors to pollution during a PCAP event, with refineries and smelters also being important contributors (Ivey et al., 2016, 2019). By using these types of results, said Ivey, it is possible to make specific recommendations about how best to reduce $PM_{2.5}$ pollution during PCAP events by targeting precursor sources.

In California, the current focus of Ivey's research, the state's Air Resources Board's estimates of the primary sources of $PM_{2.5}$ fall into the "miscellaneous processes" category, which includes residential fuel combustion, windblown fugitive dust, and managed burning and disposal. In the South Coast Air Basin,[2] such miscellaneous sources—specifically cooking and paved road dust—are again among the top sources of $PM_{2.5}$. However, said Ivey, while those sources may be the top contributors to $PM_{2.5}$ in that region, the main driver for cancer risk associated with air pollution, according to the Multiple Air Toxics Exposure Study IV (Barbosa et al., 2015), is diesel particulate matter, which does not show up as a top source of $PM_{2.5}$. Ivey added that the hot spots for cancer risk map to communities that the state identified as being at high risk of exposure to air pollution.[3]

She explained that California Assembly Bill 617[4] requires targeted air monitoring and emissions reductions in overburdened communities, such as East Los Angeles, San Bernardino/Muscoy, and the community near Wilmington and Long Beach. The biggest air pollution concerns cited by members of the San Bernardino/Muscoy community are truck idling, truck traffic, a cement plant, auto body shops, and the BNSF Railway railyard. However, the top five sources of $PM_{2.5}$ emissions identified in that community are paved road dust, mineral processes,[5] and off-road equipment—three source categories that reflect the community's concerns—as well as

---

[1] "Secondary" particles are those formed from chemical reactions in the atmosphere.
[2] The South Coast Air Basin includes the western portions of Riverside and San Bernardino Counties, the southern two-thirds of Los Angeles County, and all of Orange County.
[3] *Southeast Los Angeles Community Emissions Reduction Program Staff Report*. 2020. Los Angeles: California Air Resources Board. https://ww2.arb.ca.gov/sites/default/files/2021-04/SELA_CERP_Staff_Report.pdf (accessed August 21, 2021).
[4] AB 617; C. Garcia, Chapter 136, Statutes of 2017
[5] Mineral processes include the production of crushed rock, diatomaceous earth processing, asphalt and cement concrete production, and limestone processing.

cooking, residential fuel combustion, and light-duty passenger car emissions (Figure 2-2).

Possible Sources of Bias

Ivey then discussed biases that can confound the results of models that estimate human exposures. Spatial and temporal biases, for example, can confound acute exposure estimates. She noted that models of ozone exposure in the South Coast Air Basin tend to overestimate exposures in the morning and underestimate them in the afternoon. Another major source of bias is the so-called "change of support problem," which may arise when inferring values of a variable at places different from where those values are observed (Gelfand et al., 2001; Rongerude and Haddad, 2016).

A further issue with modeling is that the exposure assessment methods that are commonly used by researchers do not address the indoor dynamics of exposure (Goldstein et al., 2021). For example, very small and very large ambient particles are removed as they cross the building envelope, changing the size distribution of indoor particulate matter. As a result, indoor particulate matter includes more particles that tend to linger in the air longer. Overall, estimates of the infiltration factor[6] for $PM_{2.5}$ range between 0.3 and 0.8. Ivey noted that global mortality studies often neglect the role of buildings as attenuators and modulators of exposure.

Human mobility can also confound exposure estimates, with people who are more mobile at higher likelihood of having their $PM_{2.5}$ exposure misclassified when using a home-based modeling method versus a call-detail record that relies on cellular phone data (Yu et al., 2020). Ivey added that Google Maps' location history feature can also serve as a tool for characterizing time and activity patterns (Yu et al., 2019), though doing so for exposure purposes requires consent from the tracked individuals.

To study the extent to which human mobility affects disparities in personal exposure to $PM_{2.5}$, Ivey and her collaborators conducted a pilot community-based participatory research field campaign in spring 2019. The 18 adult participants in this study—all residents of the Moreno Valley, Redlands, Riverside, San Bernardino, and Yucaipa communities in inland Southern California, with San Bernardino serving as the disadvantaged community (Do et al., 2021)—wore fanny packs for 7 days; the fanny packs contained monitors that measured $PM_{2.5}$ levels every 15 seconds and a GPS data logger that recorded their location every 5 seconds. The major finding from this study was that the San Bernardino residents were more at risk from exposures in their homes compared to residents of the Riverside

---

[6] The infiltration factor is the ratio of indoor to outdoor PM, considering outdoor sources only. The measure is discussed in detail later in this chapter.

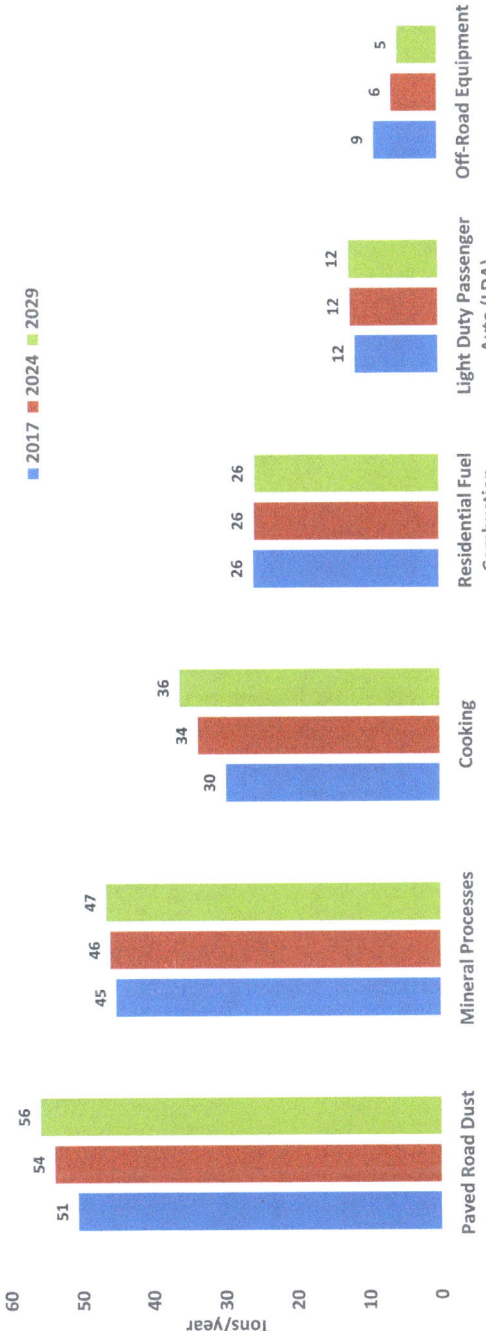

FIGURE 2-2 Top five sources of $PM_{2.5}$ emissions in the San Bernardino and Muscoy, California, community, actual and projected.
SOURCE: Ivey slide 11 (AB 617 Technical Advisory Group Meeting, July 18, 2019, slide 38).

and Redlands communities, where the poverty rate is about half that of the San Bernardino community. This finding, Ivey said, supports the idea that "when it comes to mitigation, we do need to consider people's personal behavior."

Ivey also pointed out that exposures to what she calls high-risk, non-residential points of interest should be monitored continuously, given that the participants in the pilot study spent some 30 percent of their time away from home. This work, she added, builds a foundation for using such research methods for larger scale exposure assessments, particularly those facilitated by wearable sensors as well as GPS data loggers.

Emerging Challenges

For source mitigation, Ivey noted that "we may have to consider a new paradigm for how we want to address $PM_{2.5}$ exposure disparities and therefore mitigation. We will probably have to consider these approaches through a historical lens where we consider the inequities that may confound our approaches, and we can do this by classifying these exposure disparities as either intentional or unintentional." By unintentional exposure disparity, she meant exposure that is not affected directly by humans, such as the wildfires that are emerging as an important source of air pollution as the climate changes. In this case, mitigation could involve increasing the air tightness of the building envelope and installing portable air cleaners, a subject discussed in the third session of the workshop.

What drives intentional exposure disparities are inequities in land use, said Ivey. In the Inland Empire region of California, for example, several residential communities are next to warehouses and refueling facilities for the Los Angeles natural gas-powered bus fleet. She suggested that mitigating inequities in source-related exposures may require legislation to ensure that disadvantaged communities no longer bear a disproportionate burden of pollution and environmental hazards by way of land use inequity.

To conclude her presentation, Ivey provided four takeaway messages:

1. Air pollution modeling has paved the way for understanding source-receptor relationships and associated health impacts at the population scale.
2. Misclassification and confounders complicate understanding of exposure disparities and implications for health effects and mitigation for underserved communities.
3. Building envelopes are critically important to consider for future exposure-health assessments that rely on ambient source impact data.

4. Mitigation of exposure disparities will require a commitment to sustainable and equitable zoning and development.

OUTDOOR-TO-INDOOR TRANSPORT MECHANISMS AND PARTICLE PENETRATION FOR FINE PARTICULATE MATTER

The big picture for outdoor $PM_{2.5}$ monitoring comes from the extensive network of monitors that the US Environmental Protection Agency (EPA) runs, said Brent Stephens (Figure 2-3). These monitoring stations provide accurate data for local sites and to some extent on a regional scale, but there are gaps between the stations. To fill in those gaps, researchers have developed models using personal-level data of the sort that Ivey discussed, he observed, with remote sensing data also adding to the picture (Figure 2-4). To fill local gaps, researchers are using highly local monitoring with low-cost sensors capable of providing data on outdoor $PM_{2.5}$ concentrations at the single-house scale (Bi et al., 2020; Chen et al., 2020).

As has been noted, though, Americans spend most of their time indoors, and the few studies that have examined the relationship between outdoor and indoor $PM_{2.5}$ levels have found only weak correlations between the two levels (Klepeis et al., 2001; Meng et al., 2005). As Stephens explained, nearly all outdoor air pollution epidemiologic studies fail to account for these facts, leading to what he called exposure misclassification, "where we are not quite sure if a central site or even a highly localized outdoor concentration of particulate matter is a reasonable surrogate for exposure."

Sampling indoor and outdoor particulate matter simultaneously yields an indoor/outdoor ratio, where the indoor concentration is influenced by both what is emitted or generated indoors and what infiltrates and persists from outdoors. Outdoor particulate matter infiltrates the indoor environment through open windows and other larger openings, as well as through cracks and gaps in the building envelope and mechanical ventilation (Figure 2-5). The infiltration factor, a number between 0 and 1, may be thought of as a measure that depends on eliminating the presence of indoor sources; it results in an estimate that characterizes how well a building buffers against outdoor pollution.

Infiltration Factor

Several underlying mechanisms govern the infiltration factor for any given building, said Stephens. At the highest level, the infiltration factor is a combination of what penetrates from the outdoors, a function known as the penetration factor (Liu and Nazaroff, 2001), the air exchange rate between the inside and outside of the building, and removal of particles by deposition to surfaces, phase changes (from particle to gas, for example), and air

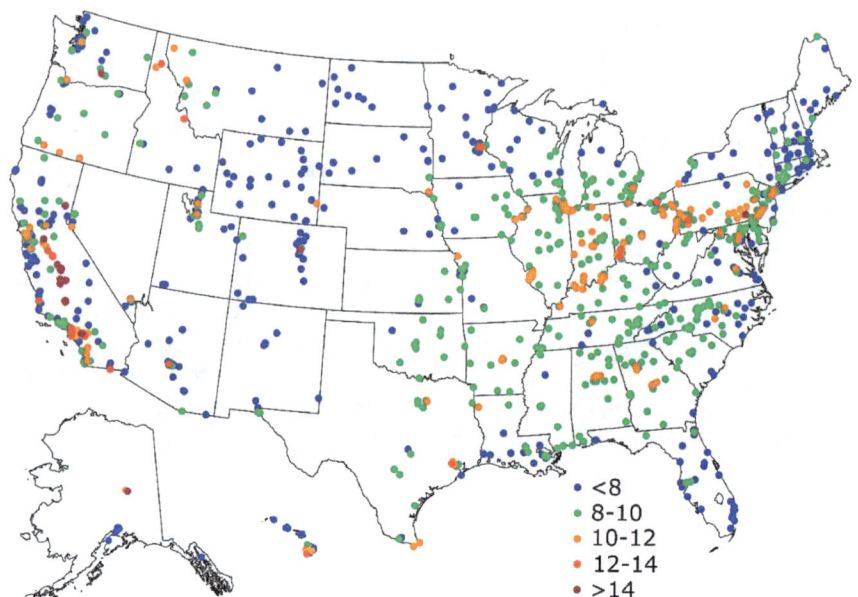

FIGURE 2-3 3-year (2013–15) average of the 24-hour $PM_{2.5}$ concentrations for the United States network of outdoor $PM_{2.5}$ monitoring stations. Units: µg/m³.
SOURCE: Stephens slide 2 (excerpted from US EPA Integrated Science Assessment (ISA) for Particulate Matter (Final Report, Dec. 2019), Figure 2-13).

filters or cleaners. The insulation in a home's walls can act as a filter that removes particles. If a building has particle penetration via mechanical ventilation, as might be the case in a large commercial building, penetration is in part a function of the type of filter the system uses.

The metrics researchers have used to measure particulate matter infiltration factors fall into two main categories, said Stephens. The most common method uses a chemical surrogate of the samples collected on a filter and weighed. Sulfur and sulfate aerosols, for which there are few indoor sources, are a reasonable proxy for the infiltration factor for various classes of particulate matter, including $PM_{2.5}$ (Sarnat et al., 2002; Wallace and Williams, 2005), though in some areas outdoor sulfate concentrations have decreased enough that it is becoming less useful as a surrogate.

Time-resolved monitoring with subsequent data processing is another approach for estimating particulate matter infiltration rates. "The idea here is if you measure indoor and outdoor concentrations of a particulate matter of some kind on a time-resolved basis, you will get data showing where periodic episodic indoor peaks shoot up over the background,"

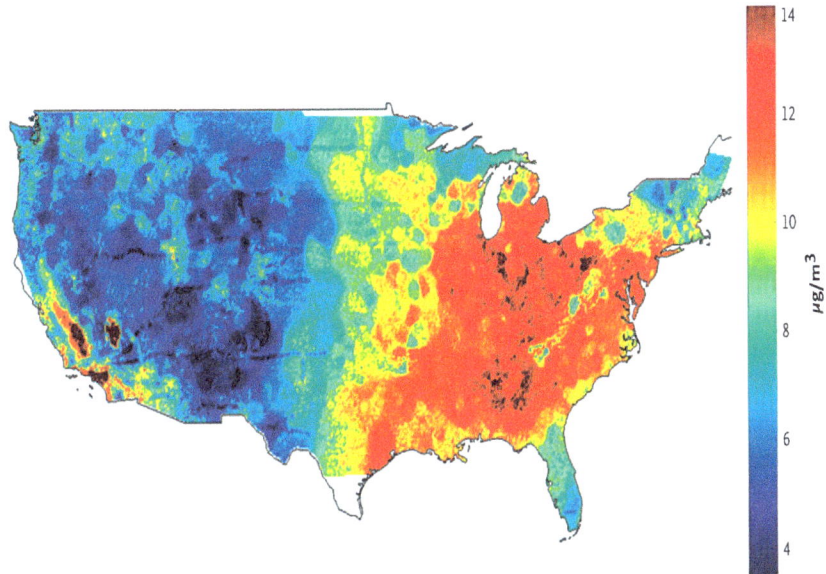

FIGURE 2-4 Average $PM_{2.5}$ concentrations on an annual basis, 2000–12, estimated by combining modeling and remote sensing information.
SOURCE: Stephens slide 3, excerpted from Figure 1A in the cited publication.[1]

[1] From *New England Journal of Medicine*. Di Q, Wang Y, Zanobetti A, Wang Y, Koutrakis P, Choirat C, Dominici F, Schwartz JD. Air pollution and mortality in the Medicare population. 376(26):2513–22. Copyright © (2017) Massachusetts Medical Society. Reprinted with permission from Massachusetts Medical Society.

Stephens explained (Kearney et al., 2014; Kunkel et al., 2017; Liu et al., 2019). Identifying the peaks and eliminating their indoor sources provides a means of comparing the indoor/outdoor ratio when there are no emissions indoors. He and a colleague had access to unoccupied apartments where there were with no indoor sources, and they found that very small and very large particles do not penetrate and persist in the indoor environment (Zhao and Stephens, 2017).

Stephens noted that a large survey of studies found that the median infiltration factor for $PM_{2.5}$ is approximately 0.5, meaning that the average building filters out about half of the outdoor $PM_{2.5}$ (Chen and Zhao, 2011). In comparison, the median infiltration factor for PM_{10} and ultrafine particulate matter ($PM_{0.1}$) is approximately 0.3 (Kearney et al., 2014; Stephens, 2015). The infiltration factor of any one home can vary tremendously, he added. The key drivers of variability in infiltration factors include pollutant characteristics such as size, class, and chemical components of the particulate matter; the source of ventilation air; human behaviors such as

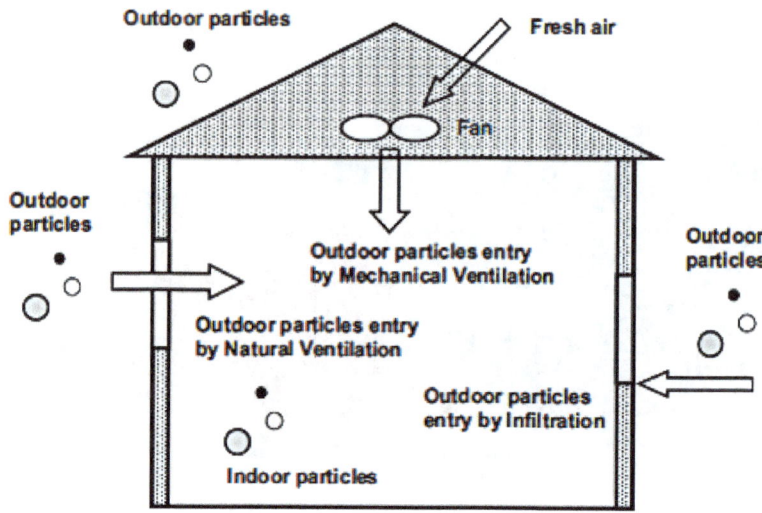

FIGURE 2-5 Infiltration of outdoor particulate matter into the indoor environment.
SOURCE: Stephens slide 6. Reprinted from *Atmospheric Environment* 45(2), Chen C, Zhao B, Review of relationship between indoor and outdoor particles: I/O ratio, infiltration factor and penetration factor, pp. 275–88, Figure 1, with permission from Elsevier.

how often the residents open and close windows and doors and their use of portable air cleaners; the magnitude of the air change rate; and the heating, ventilation, and air conditioning (HVAC) system runtime and filter efficiency.

Penetration Factors

A variety of approaches, Stephens continued, can be used to indirectly measure penetration through the building envelope. The most common methods use integrated gravimetric $PM_{2.5}$ samples (those collected on some filter medium and then weighed) and a regression analysis across homes, but there are accuracy challenges due to the lack of dynamic data. One study, for example, estimated the penetration factor for homes to be approximately 0.9 (Meng et al., 2005)—with a range of about one order of magnitude.

There is also a robust and growing literature on methods intended to measure penetration factors directly (Peng et al., 2020; Rim et al., 2010; Vette et al., 2001). In one study, researchers elevated particle concentrations in an unoccupied building, let the levels decay to background levels, used a portable HEPA (high-efficiency particulate air) cleaner to scrub the air, and then let the concentration level rebound (Thatcher et al., 2003). This approach provides a means to determine penetration factor and deposition loss rates simultaneously, but it is time-consuming and requires an empty indoor space. Stephens said he is fortunate that the Illinois Institute of Technology maintains an apartment on campus that he and his team can use for these types of studies. Together, the studies confirm that ultrafine and large particles do not penetrate and persist for long time periods in buildings.

New Directions in Assessing Infiltration and Penetration Factors

An exciting piece of research Stephens has seen recently used a low-cost sensor network to estimate infiltration factors at the scale of the building stock (Bi et al., 2021). This study used nearby outdoor PurpleAir monitors[7] matched with indoor measurements to develop a distribution of estimated infiltration factors. While this approach needs to be corroborated with data from more established measurement approaches, it is promising. Another new approach uses data from hundreds to thousands of homes outfitted with low-cost sensors to build models based on building stock characteristics, land use, geographical information system parameters, and other factors that can predict infiltrations factors across the building stock (Tang et al., 2018).

The results of these methods, said Stephens, could enable integration of infiltration factors in exposure estimates and ultimately in environmental epidemiology. He and his collaborators, for example, have examined the typical concentration response function for outdoor $PM_{2.5}$ and modified it for the underlying exposures to microenvironmental $PM_{2.5}$ exposure contributions, both indoor and outdoor (Azimi and Stephens, 2020). They are also examining how to better attribute indoor and outdoor sources across the various microenvironments in which people spend their time. So far, this approach has shown that the single largest source, on average, of $PM_{2.5}$ exposure in the population is (by a small margin) outdoor $PM_{2.5}$ that has penetrated inside homes; $PM_{2.5}$ of indoor origin is the next largest contributor.

A group at Lawrence Berkeley National Laboratory has been studying the effects of mechanical ventilation and filtration on infiltration factors in a single unoccupied residence (Singer et al., 2017a). One finding by this group, Stephens said, was that improving filtration in the house reduces the infiltration factor. However, even with a "good" filter in place, mechanical ventilation that brings in outdoor air will increase the infiltration factor and the concentration of outdoor particles inside the house. "Improving

---

[7] PurpleAir is a manufacturer of air quality monitors. The devices may be linked to a network that displays real-time measurements on a publicly accessible website (https://www.purpleair.com/map).

that supply filter then gets about the lowest infiltration factor you can," he observed. "This is not rocket science, but it is nice to have some quantitative information on how we can adjust the building stock and what certain retrofits may or may not do to the proportion of outdoor particles that penetrate and persist indoors." The findings are intuitive, but they do point to approaches that could be used to reduce $PM_{2.5}$ exposure indoors.

Over the past few years, investigators have been using health-relevant metrics in assessing infiltration and penetration factors. Stephens mentioned two studies that have examined the oxidative potential[8] of indoor infiltrated $PM_{2.5}$: one in an unoccupied test house (Khurshid et al., 2019) and the other in an unoccupied apartment (Zeng et al., 2021). These studies show that the oxidative potential of particles that infiltrate from outdoors increases indoors, and that this phenomenon is positively correlated with differences in outdoor and indoor temperature and relative humidity.

Stephens concluded his presentation with a list of research needs, which included these:

- Integrate indoor exposure attributions to ambient particulate matter epidemiology investigations to address exposure misclassification and improve the accuracy of health effect estimates.
- Determine the differential toxicity of $PM_{2.5}$ of indoor and outdoor origin.
- Address the problem that direct measurements of infiltration factors is expensive at scale. Today, such studies are typically limited to samples of convenience, but there is the potential to leverage advances in low-cost sensors.
- Improve the direct measurement of penetration factors by increasing sample sizes and incorporating $PM_{2.5}$ chemical composition, standardizing approaches, and exploring influencing factors.

OUTDOOR PARTICULATE MATTER SOURCES AND THE CHEMICAL TRANSFORMATIONS THAT TAKE PLACE WHEN THEY INTERACT WITH THE INDOOR ENVIRONMENT

Building on the foundations laid by Ivey and Stephens, Delphine Farmer focused on sources of particulate matter in the outdoor environment and what happens to those particles after they infiltrate the indoor environment. Her perspective on particulate matter is to think about it in terms of sources and sinks. She explained that the change in concentration of whatever a

---

[8] Oxidative potential is "the capability of particles to generate reactive oxygen species in a biologically relevant system, [which] may be a useful indicator of the intrinsic toxicity of PM" (Zeng et al., 2021, p. 2).

person breathes as a function of time is equal to the amount produced through direct emissions plus chemical sources arising from reactions in the atmosphere minus the amount lost through deposition and through chemical reactions, and a meteorology term that accounts for the effects of air movement (Emerson et al., 2020; Lee et al., 2013; Li et al., 2020).

Farmer also noted that while deposition is what controls airborne particle lifetime, it is arguably the least understood component in both indoor and outdoor environments. What is known is that for $PM_{2.5}$ and smaller particles, the average atmospheric lifetime is about 7 days. "What that means is that what we breathe is a combination of both local sources and sources transported from other regions," said Farmer (Figure 2-6).

Research demonstrates that particulate matter levels have been decreasing across the United States thanks to the Clean Air Act and regulatory activities that have reduced not only primary emissions but also the atmospheric chemical reactions that produce particulate matter (Jaffe et al., 2020; McClure and Jaffe, 2018). The exception is in the Western United States, where wildfire smoke is increasing the amount of outdoor particulate matter that is primarily organic in composition. This increase in organic aerosol from biomass burning, said Farmer, is influencing both outdoor and indoor air quality.

Several years ago, Farmer and her collaborators put an aerosol mass spectrometer on an airplane, flew it into wildfire plumes, and then traced the smoke plume as it diluted to study the chemistry occurring in the

FIGURE 2-6 Output from a NASA model showing the different sources of particulate matter. Red = dust, blue = sea salt, green = smoke, and white = sulfate.
SOURCE: Farmer slide 3 (NASA Center for Climate Simulation at Goddard Space Flight Center).

atmosphere (Garofalo et al., 2019; Lindaas et al., 2021; Palm et al., 2020). The first thing that happens is that the plume dilutes in a manner that affects the equilibrium of the smoke constituents. "The organic aerosol that dominates biomass burning emissions has some semivolatile components, and when you move it from a high-concentration region to a low concentration, equilibrium drives some of those semivolatile components off in dilution-driven evaporation," Farmer explained. In addition, the trace gases in wildfire smoke get oxidized, and the oxidized volatile organic compounds (VOCs) can have a low enough volatility that they condense and partition to the particle mass to form a secondary organic aerosol.

Furthermore, the volatile components released by dilution-driven evaporation can also oxidize and then condense. In fact, analysis of the trace gas and particle data collected during the project revealed that within a few hours after emission, a third of the particulate matter from smoke was actually secondary organic aerosol formed through this latter process (Palm et al., 2020). "This should tell you that the atmosphere is a dynamic environment, with many chemical reactions, and this is constantly changing the composition, the amount, and the size distribution of the particles that are in the atmosphere," said Farmer.

The sources of primary particulate matter and precursors to secondary outdoor particulate matter include anthropogenic sources—fossil fuel combustion, for example—and natural sources such as volcanic emissions and sea spray. The biosphere, Farmer added, is a massive source of VOCs that can react in the atmosphere to produce particulate matter. These sources are not evenly distributed across the planet—there is more sea spray aerosol along the coasts than in the Rocky Mountains, to cite an obvious example—so there is spatial heterogeneity in the composition of outdoor particulate matter. One recent study found that particle concentration is driven by crustal material, secondary inorganic aerosols, and biogenic secondary organic aerosol, and that anthropogenic aerosol from sources such as wood burning and vehicle wear drives the oxidative potential in that aerosol (Daellenbach et al., 2020). Oxidative potential, said Farmer, defines how reactive an aerosol is when it interacts with a cell in the human body and the harm it can produce via oxidative stress. "What this study highlights to me is that we need to understand more than just aerosol sources," she said. "We need to think about the chemistry and use the chemistry to link to health effects."

Chemistry and Indoor Particulate Matter

As both Ivey and Stephens noted, most people spend the majority of their time indoors, making it important to understand what happens to the chemistry of outdoor particulate matter as it moves indoors. The indoor environment, said Farmer, has a very high surface area and a large array of volatile compounds compared to the outdoor environment, but very low oxidant loadings. As noted, particles may remain airborne for 7 days outdoors in the absence of influences, but with mechanically ventilated spaces indoors that time shrinks to hours or minutes. In addition, once particles infiltrate the indoor environment, they exist in much lower concentrations, and the more volatile components have lower indoor-to-outdoor ratios (Johnson et al., 2017).

Temperature gradients are an additional factor to consider because particles undergo thermal partitioning as they infiltrate, Farmer observed. When they warm during infiltration they can volatilize, whereas when they move from a warm outside to a cold inside, they can condense. In fact, there are seasonal differences in the indoor-to-outdoor ratios of various VOCs in the particle phase (Avery et al., 2019b). Chemistry comes into play when considering that both human activity indoors and the indoor built environment add a huge amount of volatile compounds indoors. Oxidant levels are low indoors, but they are not zero and the oxidants that are present can react with the organic load to produce an array of oxidized organic and inorganic products that can then condense to form secondary organic aerosol (Avery et al., 2019a).

In one experiment, investigators showed that limonene, a key ingredient of anything that smells like citrus, reacts with household bleach vapors to form new particles (Wang et al., 2019). While this was a laboratory study, Farmer and her colleagues were able to reproduce that finding in a field study in which they showed that the VOCs released during cooking could react with bleach used in household cleaning to produce oxidized organic aerosol (Mattila et al., 2020). She noted that the fraction of particulate matter produced was small, which she believes is driven by the short timescales over which chemistry can occur in indoor environments.

There are many other examples of research demonstrating the important role of indoor chemistry in particle formation. Farmer cited a study in which investigators took particles of different compositions, put them in a room with vinyl flooring—which releases very small amounts of phthalate esters—and identified chemical reactions that could produce aerosols (Eriksson et al., 2020). In addition, this study found that the phthalate esters condensed more efficiently on some particles than others, showing not only that chemistry matters in the indoor environment but that pollutants can move from one place in a house to another. Another study showed that cigarette smoke generated outside can infiltrate indoors, deposit on surfaces, undergo chemical processing on those surfaces, and revolatilize on timescales of months. That so-called "third-hand smoke" can then partition onto different types of particles (Collins et al., 2018; DeCarlo et al., 2018).

In summary, Farmer said that sources of outdoor particulate matter are spatially and temporally variable, and that predicting outdoor, and thus indoor, particulate matter requires understanding emissions, chemistry, and deposition rates. In addition, indoor particulate matter undergoes equilibrium partitioning in terms of dilution and temperature as well as chemical reactions. Given that, she concluded with three open questions and research needs:

1. What is the indoor budget of condensable material (that is, its concentration, reactivity, and volatility)?
2. Which multiphase reactions compete with ventilation and deposition and over what timescales?
3. How do particle composition and phase affect indoor particulate matter chemistry?

Her conclusions on the basis of the studies discussed were that there is broad scientific consensus that outdoor $PM_{2.5}$ influences indoor $PM_{2.5}$, so reducing outdoor sources improves indoor air quality; and that once indoors, aerosols have many chemical fates—albeit on far shorter timescales than outdoor air—and that the resulting chemical composition of the air may influence human health.

DISCUSSION

Kimberly Prather, the moderator, invited questions from planning committee members and then from the webinar viewers.

Elizabeth Matsui noted that she was struck by the complexity of the outdoor sources, the penetration of the indoor emissions from them, and the chemistry involved in creating particulate matter. She then asked the speakers if there is a framework to identify and prioritize the best targets for mitigation. Farmer replied that while the outdoor atmosphere is complex, many of the underlying fundamental principles are understood, including about the chemistry occurring on regional and global scales. The open questions, she said, are more at the local level that Ivey addressed in her presentation. Though applying those underlying chemical and physical processes to indoor environments will be challenging, it is most likely not an intractable problem.

Stephens said that there is some evidence that factors such as air conditioning usage modifies some of the outdoor air epidemiology associations with $PM_{2.5}$, suggesting that filtration and newer building stock can mitigate some of the adverse health effects. He quoted a mantra in building science, "build tight, ventilate right," but said it requires efficient filters to prevent buildup of particulate matter, and noted that the marketplace is not meeting the need for effective and energy-efficient ventilation for homes. "None of this is rocket science," said Stephens. "It is about priorities and incentives." Addressing the existing building stock is a challenge, he added, in part because it is not well understood.

Rengie Chan asked the panel if there is a need for more research on the efficacy of sealing homes to reduce infiltration and how to translate that into practical mitigation steps. Stephens answered that unpublished data from his group showed that energy retrofits did not have a big effect on infiltration factors in part because of varying meteorological conditions and variability in air sealing. But he added that if he had to make a decision, he would say sealing homes would have a big effect, although "exquisite" data do not exist showing that to be true.

Prather observed that sealing buildings goes against all the advice regarding COVID-19—to "ventilate, ventilate, ventilate"—and she asked the panel to talk about trade-offs of reducing exposure to outdoor air pollutants and increasing indoor exposure to viruses. Stephens responded that good filtration is the key to balancing those two concerns. Farmer noted that the key was to seal a building well *and* ventilate it properly.

With regard to marginalized and underserved communities, Richard Corsi asked the speakers to comment on the ability of such communities to reduce their exposures when they may not be able to afford to purchase more expensive filters or good, portable HEPA cleaners. Ivey replied that the nation needs to get creative about how to leverage policy and perhaps use programs such as Medicare and Medicaid to cover the interventions that would improve health. "This needs to become a systemic intervention because the causes of disproportionate exposures are systemic," said Ivey. "They are not individualized causes, and to fix a systemic issue, you need systemic solutions." Farmer added that the nation needs to think about the different sources of outdoor particulate matter, which are out of an individual's control, and understand what those sources are and what they contribute to the health burden in disadvantaged communities. In terms of the many opportunities that exist to decrease particulate matter exposure and improve indoor air quality, she noted that while this is a socioeconomic issue, there is a need for the scientific community to do a better job explaining to all communities which approaches are the best for producing the desired results in a cost-effective manner.

Several online participants wanted the panelists to identify what the best metric would be for quantifying acceptable levels of indoor particulate matter. Farmer explained that the challenge of developing standards is having enough information to know what levels people are exposed to, what sources one can actually control, and then what the health effects are of the various constituents of indoor particulate matter. For ultrafine particles, there is a growing body of evidence that they can penetrate deep into the lungs and produce adverse health effects. In her view, many of the questions about regulations or acceptable levels should be guided by the precautionary principle, the idea of first figuring out that certain levels are safe and being cautious about setting guidelines. "The science is there. It is very clear that it shows that particulate matter has negative consequences to human health," said Farmer. She added that the science is also clear that there are dangerous compounds in the air that can be produced by chemistry and that some have negative health effects. "I think we can use that information to start to develop guidance," she said.

Stephens noted that $PM_{2.5}$ is often used as a blanket measure because it is harder to get data on composition and size resolution for smaller particles. Epidemiology studies, in particular, struggle in that regard, in part because of the way monitoring and regulatory frameworks have been established. He noted a study from the Center for Nanotechnology and Nanotoxicology at the Harvard T.H. Chan School of Public Health showing that particles generated by photocopiers and printers at a copy center had the potential to harm the lungs of those exposed to them (Pirela et al., 2013). While he characterized the framework for this study as elegant, with the potential to increase understanding of the competing effects of indoor and outdoor sources, he added that it took tremendous resources to conduct that one study at one copy center.

Ivey remarked that there is a need to think about where the greatest disparities exist and to consider who is most negatively affected and how they are most negatively affected, given that society is only as strong as its "weakest" population. One step would be to conduct more surveys in the communities that have the highest exposures and that are surrounded by the sources that potentially lead to cancers, and then establish standards to protect those who are most vulnerable in society. In the same vein, Farmer said that there is a great opportunity to apply the technologies developed to study and understand the association between what happens in the outdoor environment and the indoor environment and do so in the most vulnerable and disadvantaged communities.

An online participant wanted to know what criteria the research community is focused on to quantify the intersection between particle infiltration and health, particularly regarding time-integrated exposure versus instantaneous maximum exposure. Stephens said the literature shows that both are being studied. The most robust associations to build on regarding health outcomes are those on long-term exposures to particulate matter, with few studies on the daily exposure component. He added that there have been only a handful of studies that examine whether infiltration affects emergency room visits, for example. Prather concurred with that assessment, noting that spikes in outdoor particle levels could have important health effects but there have been few studies examining that idea, and

Farmer said the challenge arises because indoor environments may have high ventilation rates and indoor concentrations can change rapidly. This can result in circumstances where there are repeated short-term, high-concentration exposures. "I do not think we really understand how those different timescales of exposure link to health effects," said Farmer.

The final question to the panel asked if there are any studies on long-term (50- to 80-year) trends in air quality versus human health, wondering if at this point only incremental benefits are being achieved from further, potentially expensive interventions. Stephens said that a 2009 paper described changes in indoor air pollutants since the 1950s (Weschler, 2009), but he was not aware of much more than that. Farmer noted that there are good data on long-term trends for outdoor particulate matter and clear links to changes in hospital visits, mortality rates, and cardiorespiratory disease. The data show that reducing particulate matter exposure by even small amounts has very substantial beneficial health effects, which is why the Clean Air Act is viewed as one of the strongest pieces of legislation in terms of extending life years of people in this country.

3

Indoor Sources of Indoor Particulate Matter

While the workshop's first session focused on outdoor sources of indoor particulate matter, the second featured four presentations on indoor sources of indoor particulate matter. Marina Vance (University of Colorado Boulder) discussed the role of cooking in generating fine particulate matter ($PM_{2.5}$). Michael Waring (Drexel University College of Engineering) described how secondary aerosols form $PM_{2.5}$, and Linsey Marr (Virginia Tech) talked about the effect of humidity on the chemistry and biology of indoor air. In the session's final presentation, Andrea Ferro (Clarkson University) explained how different sources of indoor $PM_{2.5}$ influence the manner in which exposure and health effects are characterized. Kimberly Prather again moderated an open discussion after the presentations.

FINE PARTICULATE MATTER EMISSIONS FROM COOKING

Cooking is an important source of indoor $PM_{2.5}$ and can vary dramatically from one home to another, said Marina Vance. As a result, the inhabitants of two adjacent homes could have very different exposures to $PM_{2.5}$. She explained that there are two components to the air pollution generated by cooking: the heat source and the food. Primary emissions from cooking can include particles, VOCs, nitrogen oxides, and other pollutants.

Heat sources can include solid fuels such as coal, wood, charcoal, and crop residues; gaseous fuels such as propane, natural gas, liquefied petroleum gas, and biogas; electricity; and induction. Vance noted that some 2.8 billion people worldwide depend on solid fuels for cooking and those fuels led to upwards of 4 million premature deaths in 2010, according to data from the Global Burden of Disease Study (Chafe et al., 2014; Smith et al., 2014). In the developed world, gas and electric heat sources are cleaner, so the emissions from food are the main source of indoor pollutants.

Cooking releases primary emissions that grow into a specific particle size and move through the indoor environment, where they can take up semivolatile organic compounds (SVOCs) from building materials and other sources, Vance explained. In addition, the volatile fraction of cooking emissions can undergo reactions indoors that may produce particles, and a large fraction of indoor particulate matter from cooking is likely to deposit on indoor surfaces. Some of these emissions may find their way outdoors, where they may undergo atmospheric processing, and the volatile components in particular could form secondary organic aerosols outdoors and impact ambient air quality as well.

Primary Emissions

The House Observations of Microbial and Environmental Chemistry (HOMEChem) study,[1] for which Vance and Delphine Farmer were the principal investigators, aimed to identify the most important aspects of the chemistry that controls the indoor environment using the University of Texas at Austin's test house, a 1200-square-foot manufactured home. One set of experiments observed particle emissions during various cooking activities (Figure 3-1) (Patel et al., 2020).

In one experiment that involved no cooking and opening and closing all the doors and windows in the house to allow maximum penetration of outdoor air pollution, indoor concentrations of coarse particulate matter (PM_{10}) predominated at levels of 5–10 micrograms per cubic meter ($\mu g/m^3$) (Figure 3-1, bottom panel). When intense cooking activities occurred with no ventilation—the so-called "Thanksgiving Day" experiment because it involved cooking a large meal meant to serve 12–16 people—there were intense peaks of all sizes of particulate matter, including $PM_{2.5}$, that soared to hundreds of $\mu g/m^3$ (Figure 3-1, top panel). Vance noted that the levels of ultrafine aerosols ($PM_{0.1}$) rose significantly at some specific times of that day, demonstrating that it is not always true that $PM_{0.1}$ accounts for very little of the particle mass in the indoor environment (Figure 3-2). In fact, at some times during this particular experiment $PM_{0.1}$ accounted for most of the $PM_{2.5}$, while at other times, $PM_{0.5}$ accounted for the majority of the total particulate mass. This finding, she said, has implications for health, exposure, and the use of low-cost sensors as a means of evaluating air quality in the indoor environment.

---

[1] Additional information is available at https://indoorchem.org/projects/HOMEChem/.

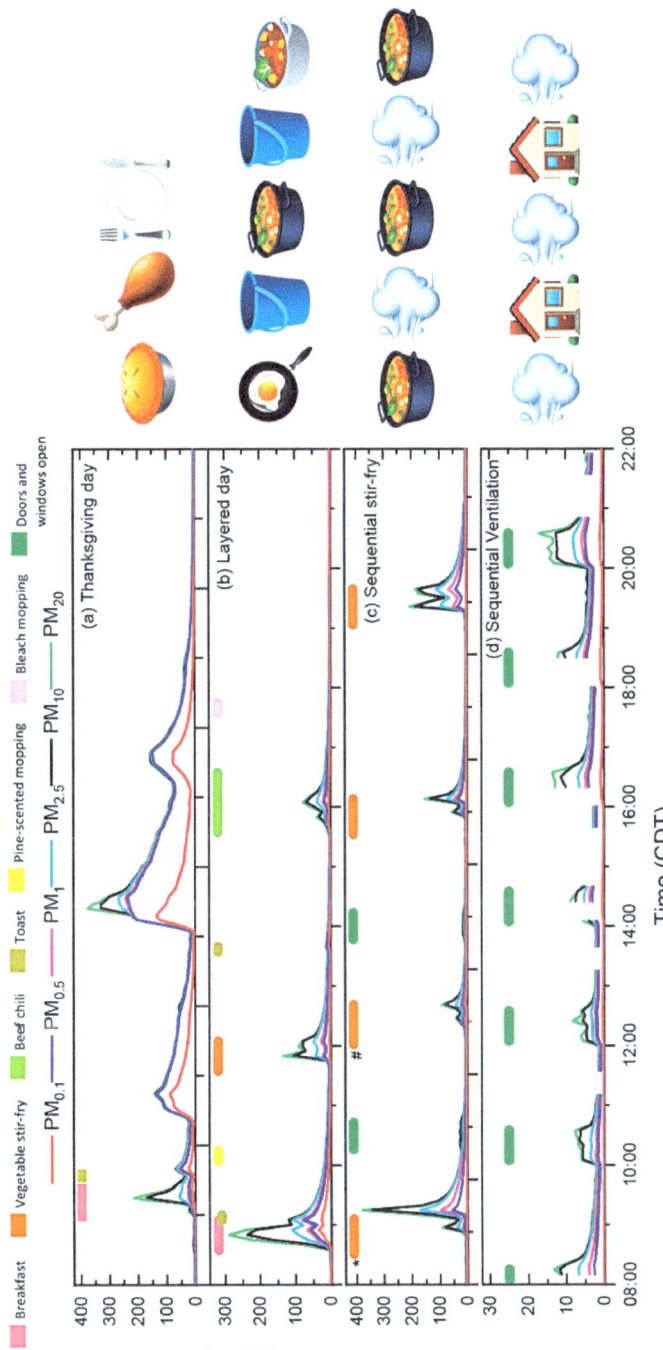

FIGURE 3-1 High particle concentrations observed during cooking activities, with and without ventilation. CDT = Central Daylight Time; PM = particulate matter.
SOURCE: Vance slide 7, adapted from Patel et al. (2020) Figure 1.[1]

[1] https://pubs.acs.org/doi/10.1021/acs.est.0c00740; reprinted with permission from American Chemical Society Publications. Further permission related to the material excerpted should be directed to the American Chemical Society.

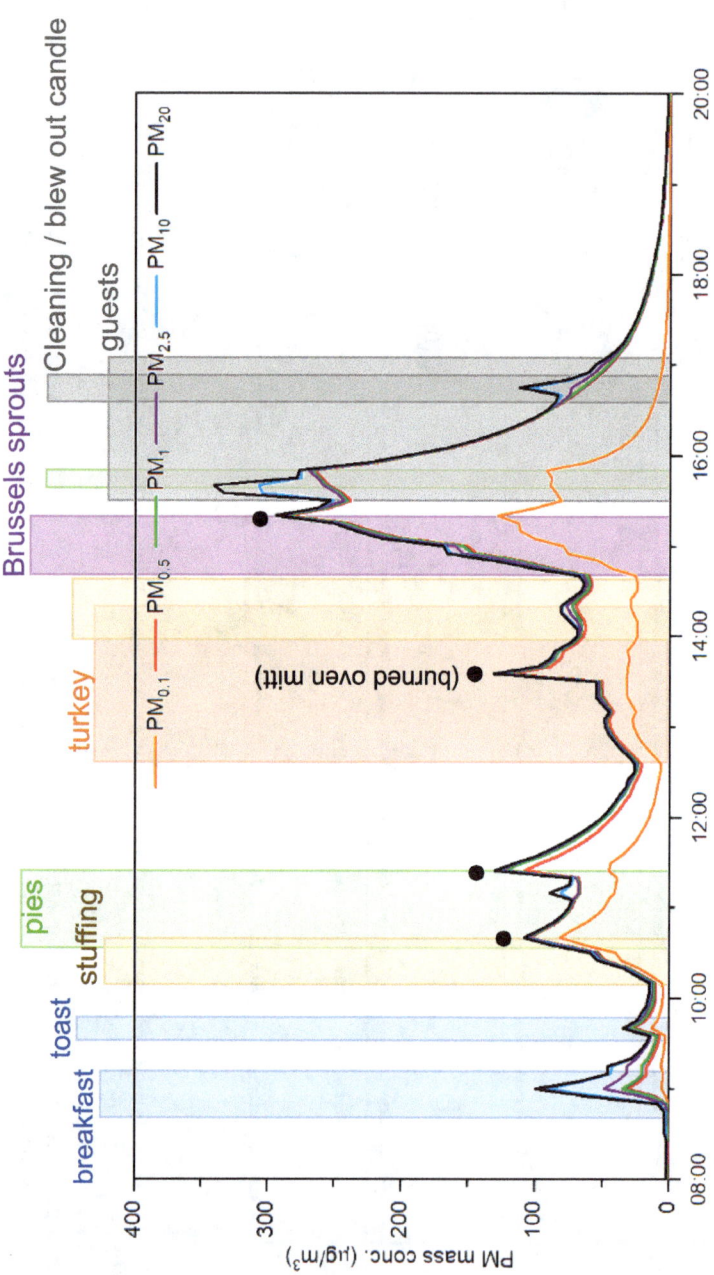

FIGURE 3-2 Total indoor particulate matter mass concentrations during the HOMEChem Thanksgiving Day experiment.
SOURCE: Vance slide 8, adapted from Patel et al. (2020) Figure S5.[1]

[1] https://pubs.acs.org/doi/10.1021/acs.est.0c00740; reprinted with permission from American Chemical Society Publications. Further permission related to the material excerpted should be directed to the American Chemical Society.

Scenarios designed to reproduce more typical cooking activities found that meals prepared on a propane stove produce approximately fourfold higher levels of particulate matter than meals cooked on an electric hot plate or in the oven or toaster. Another finding was that the heat source might be responsible for particle number levels, while the food may dominate mass concentrations.

In terms of composition, multiple studies have shown that indoor cooking produces substantial amounts of primary organic aerosol (Farmer et al., 2019; He et al., 2004; Katz et al., 2021; Klein et al., 2019), although there is a great deal of variability that could depend on what is being cooked, the cooking style, heat source, type of cooking oil, and other details (Farmer et al., 2019; Torkmahalleh et al., 2017). In particular, the concentration of polycyclic aromatic hydrocarbons associated with cooking emissions was highly variable, which can have serious implications with regard to health effects (He et al., 2004; Torkmahalleh et al., 2017).

Emitted particles can adsorb SVOCs released during cooking, but the HOMEChem experiments found that other substances, such as phthalates and adipates used in vinyl flooring and cyclic siloxanes from the oven and building materials, also ended up as part of the $PM_{2.5}$ (Lunderberg et al., 2020). A similar study conducted in a house during normal occupancy found that the total concentration of SVOCs rose when there were indoor activities going on, particularly cooking (Kristensen et al., 2019) (Figure 3-3). In fact, 14 percent of all the volatile organic compounds present during cooking, and 25 percent when the oven was in use, were in the particle phase. Vance added that the SVOCs and particles emitted during cooking may undergo chemical reactions indoors. For example, the HOMEChem study found that SVOCs emitted during cooking could remain in the air and react later with components of a bleach solution used to clean the house.

Surface deposition can serve as an important sink for indoor particulate matter rich in organic matter. HOMEChem experiments found that although the Thanksgiving Day experiment produced higher airborne particulate matter concentrations than during a normal cooking event, the amount of particles deposited on a piece of glass (which was set up for testing purposes) and their size distribution were similar (Or et al., 2020). Another analysis of the HOMEChem data showed that the material accumulated on a glass surface had similar properties to the organic aerosol generated by cooking as measured by online aerosol mass spectrometry (O'Brien et al., 2021).

These results, said Vance, indicate that the particles detected in the air ended up on surfaces in the house—and in the human respiratory system. For a person spending the day outdoors during the HOMEChem test

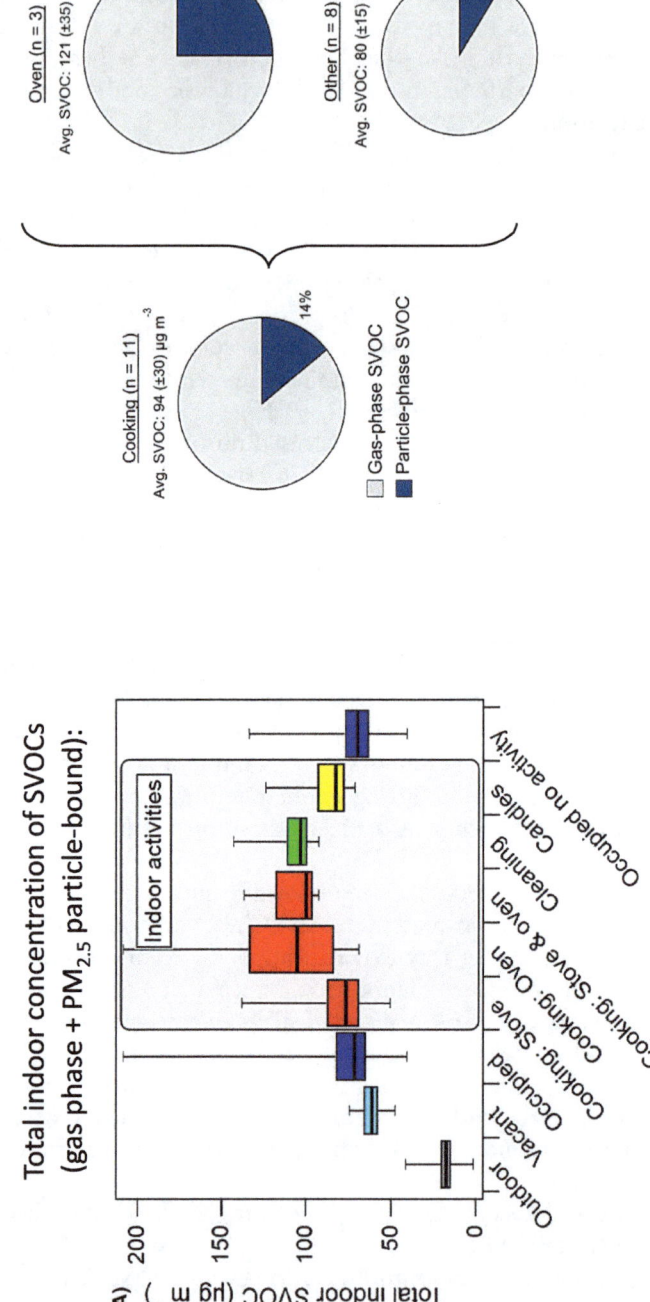

FIGURE 3-3 Indoor residential activities enhance semivolatile organic compound (SVOC) concentrations.
SOURCE: Vance slide 15, adapted from Kristensen et al. (2019) Figure 5A and Figure 7; reprinted with permission from John Wiley and Sons, Inc.

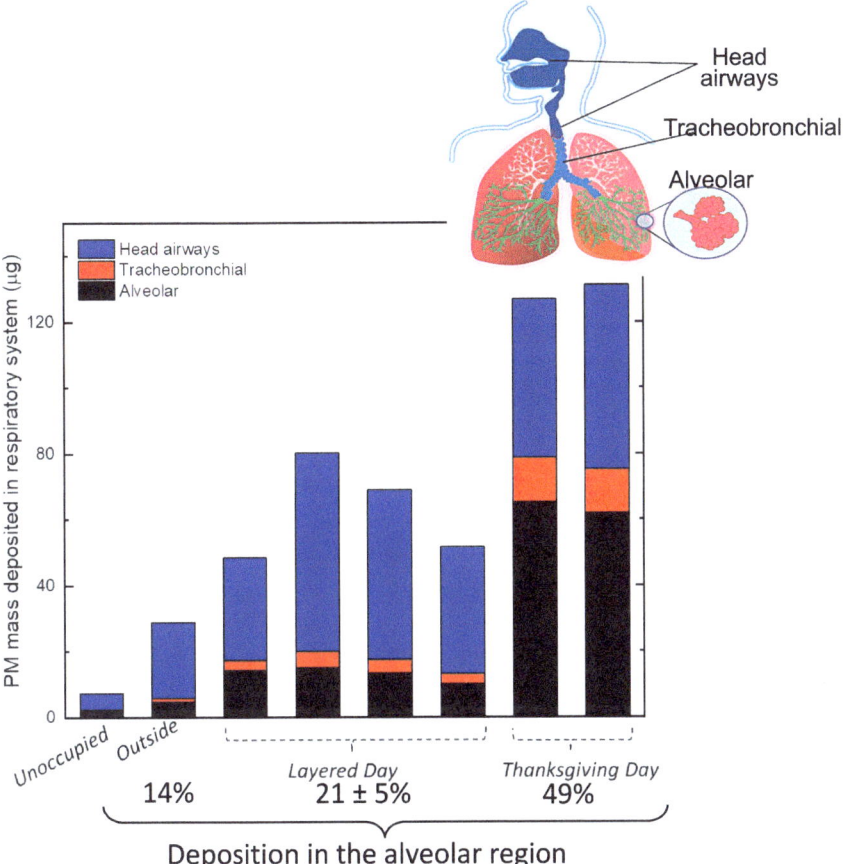

FIGURE 3-4 Particulate matter (PM) mass deposited in the respiratory system in different contexts.
SOURCE: Vance slide 21, adapted from Patel et al. (2020) Figure 4.[1]

[1] https://pubs.acs.org/doi/10.1021/acs.est.0c00740; reprinted with permission from American Chemical Society Publications. Further permission related to the material excerpted should be directed to the American Chemical Society.

period, about 14 percent of the aerosol they inhaled would end up in the alveolar section of the respiratory system (Figure 3-4) (Patel et al., 2020). However, for someone participating in the Thanksgiving Day experiment "that would be about 50 percent, so half of the particles that they are inhaling would end up in the alveolar system," she said. The reason for that increase, she noted, is because the average size distribution for a 24-hour period during HOMEChem was under 100 nanometers. She added that at least one study has shown that the toxicity of inhaled particles can vary with the way food is cooked (Alves et al., 2021).

Despite the rich data generated by the HOMEChem study and others, there remain persistent knowledge gaps related to the particulate matter generated during indoor cooking. The literature indicates questions about commonalities or differences among cooking styles and ingredients, about how ventilation could affect the fate of particulate matter and exposure, and whether the chemical composition of cooking-generated particulate matter impacts its potential health effects. In terms of chemical composition and health effects, Vance remarked, questions remain about whether the bulk composition and size are important and what the influence of sorbed compounds is on health. More work is also needed to develop metrics for particle number versus mass and dose versus exposure time as they pertain to exposure and health effects. In addition, many questions remain about the contributions of indoor cooking activities to ambient air quality and the impact of socioeconomic disparities.

Vance concluded by noting that there is broad scientific consensus that for point sources of air pollution, it is critical to control emissions at their source. "For indoor cooking emissions in that context, what that really means is to electrify cooking, because *cleaner* fuels like the gaseous fuels are not necessarily *clean* fuels," she said. For cooking, effective source control means operating a high-quality, externally vented over-the-stove hood every time cooking activities occur or using a well-positioned, properly selected portable air cleaner.

SECONDARY AEROSOL FORMATION OF FINE PARTICULATE MATTER IN THE INDOOR ENVIRONMENT

Organic aerosol, explained Michael Waring, comprises hundreds of thousands of types of molecules existing in a state of dynamic equilibrium between the gas and particle phases. Researchers classify organic aerosols based on whether they are primary aerosols produced by combustion or secondary aerosols that are the products of gas-phase reactions. Mass spectrometry is commonly used to categorize organic aerosols according to their emission type and character, with major categories including hydrocarbon-like, biomass burning, cooking, and both semivolatile oxygenated and low-volatility oxygenated (Sun et al., 2012; Zhang et al., 2011).

Most secondary organic aerosols formed indoors likely result from the reaction of ozone with an alkene such as terpene, which is emitted by building and consumer products (Kroll and Seinfeld, 2008). This reaction proceeds via a complex mechanism that generates many different products, including carbonyls, alcohols, carboxylic acids, peroxides, and organic nitrates. These tend to be lower-volatility chemicals that partition to form secondary organic aerosols, Waring said. He noted that most studies on indoor secondary organic aerosol formation have been conducted in chambers or test rooms with intentional releases of reactants (Coleman et al., 2008; Destaillats et al., 2006; Sarwar et al., 2003; Waring, 2014; Weschler and Shields, 1999; Youssefi and Waring, 2012). Early studies relied on particle counters such as scanning mobility particle sizers; more recently aerosol mass spectrometers have yielded a major improvement in measurement capability and provided compositional information on aerosols that reveal chemical trends over time. Models of indoor secondary organic aerosols have also produced important insights.

Before discussing some of this modeling work, Waring provided an example of secondary organic aerosol formation in an experiment using an air freshener operated continuously at three air changes per hour in a chamber into which ozone was piped (Destaillats et al., 2006). Almost immediately, a strong burst of nucleation occurred in which new, small secondary organic aerosols formed. These grew into larger particles via gas-to-particle partitioning mechanisms before the size distribution stabilized.

Secondary Organic Aerosol Modeling

There are two approaches to modeling secondary organic aerosols: explicit production modeling, where products are tracked explicitly, assigned partitioning equations, and allowed to partition into an existing generic organic aerosol (Carslaw et al., 2012; Kruza et al., 2020); and lumped product modeling (Cummings et al., 2020; Cummings and Waring, 2019; Youssefi and Waring, 2012), which Waring uses. Lumped product modeling has particular strengths relative to explicit product modeling in that it can account for any arbitrary type of organic aerosol source, both indoor and outdoor; is not limited to secondary organic aerosols; and allows easy computation of bulk aerosol properties such as density or hygroscopicity.

After describing the parameters and equations his group uses in its models, Waring explained that this model can yield organic aerosol concentrations based on the volatility of organic mass and oxidation state. It shows that organic mass forms and ages via two processes: functionalization, in which a molecule gains oxygen-containing functional groups that enhance particle-phase concentration, and fragmentation, in which particle-phase molecules break into lighter fragments that can evaporate into the gas phase and reduce particle-phase concentration. He noted that organic mass can enter from outdoors along with ozone and the many volatile organic compounds that the model accounts for (terpenes plus other alkenes, alkanes, aromatics, and the like) or be emitted as primary organic mass indoors along with the same VOCs and nitrous acid (HNO_2 or HONO). Losses occur via deposition on surfaces or air exchange.

Waring and his collaborators have used the lumped product model to estimate the likely magnitude of secondary organic aerosol formation from VOC reactions with ozone or hydroxyl radicals in typical residences in the United States (Waring, 2014). These simulations predict that organic aerosols account for much of the indoor particulate matter and that primary organic aerosols, largely generated by cooking, dominate the organic aerosol content. While secondary organic aerosols are typically the smallest component of indoor organic aerosols, they can dominate both the aerosols and particulate matter when ozone and terpene levels are high and air exchange rates are low.

Research Needs

In the final portion of his presentation, Waring highlighted future research needs. The first concept that needs exploration is whether low-concentration, fast-reacting stealth reactants present in indoor air could have large effects on secondary organic aerosol formation. The origin of this idea was the surprising modeling result that, while d-limonene drives most secondary organic aerosol formation, the next most important driver was α-terpinene, and while the concentration of this compound is half that of d-limonene, it reacts ten times faster.

Another research need is to investigate whether the phase state of organic aerosol influences secondary organic aerosol partitioning and whether equilibrium assumptions about partitioning indoors are valid or secondary organic aerosol formation might be kinetically limited instead. The genesis of this question, said Waring, is modeling showing that most indoor organic aerosols are semisolid rather than liquid or amorphous solid, suggesting that it might be productive to use kinetic partitioning frameworks in models rather than equilibrium partitioning.

Waring then raised the question of whether nucleation or partitioning drives indoor secondary organic aerosol formation, as it is unclear which predominates based on modeling results. He also thought it important to know whether squalene present on human skin can react with ozone to form secondary organic aerosol particles, which a simple laboratory study suggested might be possible (Wang and Waring, 2014), and to identify effective mitigation strategies for indoor secondary organic aerosol formation. "I think we should be asking ourselves which terpenes are most important to remove from indoor formulations," he said.

He also believes it would be useful to study whether standard practices are effective at removing secondary organic aerosols from indoor air, given that they are dynamic mixtures. In other words, removing some of the organic aerosol content would just lead to the system readjusting.

The final research need that Waring identified was to better understand the influence of ozone and hydroxyl-producing technologies, such as ultraviolet lamps and bipolar ionization found in newer ventilation systems, on secondary organic aerosol formation.

THE EFFECT OF HUMIDITY ON THE CHEMISTRY AND BIOLOGY OF INDOOR AIR

Humidity is not a direct source of indoor particulate matter, but it does affect the chemistry and biology of airborne substances, said Linsey Marr. Absolute humidity (an expression of the concentration of water in the local atmosphere) can range from zero to 40 grams per cubic meter. Relative humidity is the amount of water vapor in the air compared to the maximum amount that the air can hold at a specific temperature. Absolute and relative humidity are strongly correlated indoors simply because the temperature varies within a narrow range (Marr et al., 2019).

Surprisingly, said Marr, there are few data on how indoor and outdoor humidity relate to one another, and studies that have examined the health effects of humidity have relied on outdoor data. One set of studies found that indoor and outdoor absolute humidity in temperate regions are correlated during the non–air conditioning season and that indoor relative humidity is better correlated with outdoor absolute humidity than outdoor relative humidity (Nguyen and Dockery, 2016; Nguyen et al., 2014). She noted that the pattern differs in tropical regions, where indoor absolute and relative humidity are linearly correlated with outdoor humidity in the absence of air conditioning, suggesting that outdoor absolute humidity may be a good proxy of indoor humidity in the absence of air conditioning (Pan et al., 2021).

Humidifiers can be a source of particulate matter indoors, noted Marr, particularly when using tap water, which contains dissolved minerals and even undissolved solids that can be aerosolized by nebulizing or ultrasonic humidifiers (Yao et al., 2019, 2020). Deionized water produces fairly low levels of particles, while water with dissolved solids produces a large number of particles. Water hardness does not seem to have much effect.

Humidity, Marr continued, has a large effect on the viability of bacteria and viruses in the air (Lin and Marr, 2020). Bacteria survive worst at low relative humidity, while virus survival is lowest at midrange levels of relative humidity, in both aerosols and stationary droplets. Relative humidity is important for chemistry, she explained, because it controls evaporation and condensation, and hence the size of airborne particles. For example, the size of a particle containing physiologic levels of protein, salt, and influenza shrinks by more than half when relative humidity decreases from close to 100 percent to 20 percent, with most of the change taking place between near 100 to 80 percent (Marr et al., 2019; Mikhailov et al., 2004).

The weight of evaporating droplets depends on the rate of evaporation, which relative humidity controls. At low relative humidity, the weight of a droplet drops rapidly and solute[2] concentrations can rise dramatically. At high relative humidity and slow evaporation, droplets lose far less of their weight and solute concentrations change only slightly (Lin and Marr, 2020). What Marr finds interesting is that at medium relative humidity, evaporation occurs more slowly and remains incomplete (Figure 3-5). She believes this occurs because the droplets form some sort of semisolid gel.

Mold, which thrives at higher humidity levels, is thought to be a problem in 18 to 50 percent of US buildings (Mendell et al., 2011), and concern about it accounted for some 90 percent of the calls and emails Marr received before the COVID-19 pandemic. Mold releases spores in the 2 to 100 micron size range, and they can act as allergens; exposure to mold is associated with upper respiratory tract symptoms, cough, wheeze, asthma, and other health problems (Fisk et al., 2007; Mendell et al., 2011), although Marr said there has been little research on the indoor emission rates of mold spores. Mold also releases microbial VOCs, and the chemistry and toxicity of these are also greatly underexplored, she added.

The pH of a droplet likely plays an important role in both the chemistry that takes place and microbial survival, but it is challenging to measure the pH of a 50 micron or smaller droplet. To address that problem, she and a colleague developed functionalized gold nanoparticles that they seed into a suspension that can then be aerosolized. The nanoparticles have a pH-sensitive dye that can be detected with surface-enhanced Raman spectroscopy (Freedman et al., 2019). In one experiment, Marr's team found that the pH of droplets formed from water buffered at pH 7.3 was 10.5 ± 1.0, which would have a significant effect on the chemistry that could take place in those droplets. Further analysis found a spatial pH gradient in the droplets, with the highest pH values in the center of the droplet.

Marr noted that relative humidity and pH can have a significant effect on virus viability (Lin et al., 2020), demonstrating that relative humidity, which dynamically controls the water content of particulate matter and hence the concentration of the chemicals, bacteria, and viruses in those particles, affects both the chemistry and biology in aerosols.

---

[2] A solute is a component of a solution, dissolved in a separate solution (the solvent).

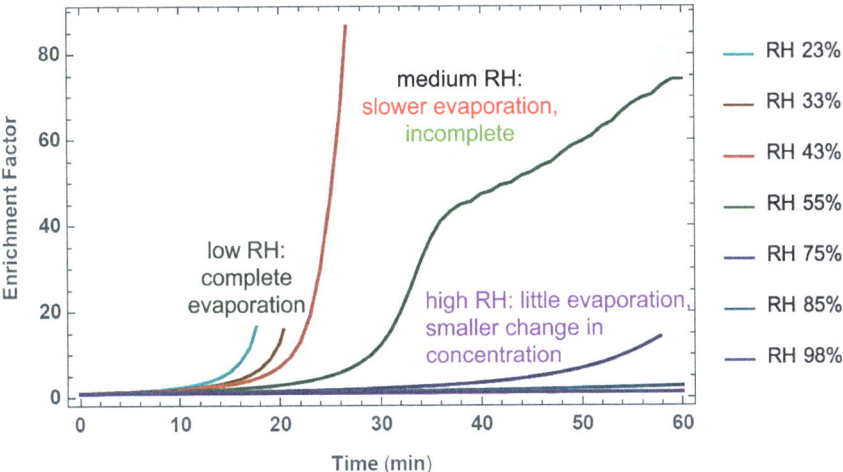

FIGURE 3-5 Change in solute concentrations at varying relative humidities (RH). SOURCE: Marr slide 9, adapted and reprinted with permission from Lin and Marr (2020) Figure 3B; Copyright 2021 American Chemical Society Publications.

THE INFLUENCE OF SOURCES OF INDOOR FINE PARTICULATE MATTER ON THE CHARACTERIZATION OF EXPOSURE AND EVALUATION OF HEALTH EFFECTS

To start the day's final presentation, Andrea Ferro highlighted a 1996 review paper on indoor particles that summarized studies in what was then a fairly new field (Wallace, 1996). At the time, the major known sources of indoor particulate matter were smoking, cooking, and "unexplained," and the author noted that exposures in the "personal cloud" (an individual's breathing zone) were higher than what was predicted from indoor monitoring. The paper also pointed out that personal $PM_{2.5}$ exposures correlated poorly with outdoor $PM_{2.5}$ levels, except when looking at individuals over long time periods.

Smoking and cooking are still considered the major sources of indoor particulate matter, and many of the "unexplained" sources are now known. Still, said Ferro, relevant research questions remain, such as the following:

1. How does the indoor environment, including sources, occupants, building operation, and pollutant/microbial cycling, affect the level and composition of indoor $PM_{2.5}$?
2. How do acute and chronic exposure to indoor $PM_{2.5}$ affect human health and how should health risks of indoor $PM_{2.5}$ be assessed?

3. What are the structural influences on disparate $PM_{2.5}$ exposures across communities?
4. What are the best approaches to mitigate indoor $PM_{2.5}$ exposures?

Ferro pointed out that levels of indoor $PM_{2.5}$ are often episodic, given the variety of sources; cooking, for example, is an episodic source, while a stove pilot light is a continuous source. And indoor and outdoor $PM_{2.5}$ temporal patterns can be quite different. Indoor patterns change throughout the day and the week, reflecting the activities of the residents, while outdoor patterns are strongly associated with sources such as the level of vehicle traffic.

Given that short-term (<24-hour) increases in outdoor $PM_{2.5}$ are associated with increased mortality (Lin et al., 2017; Lu et al., 2015) and that there is some evidence that short-term exposure to $PM_{2.5}$ affects morbidity, Ferro wondered how indoor peaks in $PM_{2.5}$ affect human health. Answering that question will require exposure estimates of indoor $PM_{2.5}$ to assess acute health effects, such as myocardial infarction, that are known to be associated with peaks of $PM_{2.5}$ concentrations, she said. In fact, indoor particulate matter may be more toxic than outdoor particulate matter. One study, for example, found that the inflammatory responses and allergic reactions triggered by PM_{10} collected at six schools in Munich were higher than for PM_{10} collected outdoors (Oeder et al., 2012). Studies using rodent macrophages found that indoor particles produced larger proinflammatory (Long et al., 2001) and cytotoxic effects (Happo et al., 2014) than outdoor particles.

The chemical content of indoor $PM_{2.5}$ is enriched by bioorganic compounds such as squalene, cholesterol, and fatty acids, as well as compounds in chemical additives such as plasticizers, flame retardants, and surfactants. Each of these constituents, said Ferro, partitions between the gas and particle phases in a manner dependent on temperature and relative humidity, as well as chemical transformations including oxidation and acid-based and partitioning reactions. Indoor particulate matter also contains microbes that may originate from humans, plants, pets, plumbing systems, HVAC equipment, mold, resuspension of settled dust, and outdoor air (Prussin and Marr, 2015). Ferro mentioned a 2017 National Academies report that summarized research on the indoor microbiome and the built environment (NASEM, 2017), and noted that information on virus transmission through aerosols has increased substantially as a result of the COVID-19 pandemic.

In terms of source apportionment, techniques such as positive matrix factorization can determine the sources for human exposure, including indoor exposure. One study, for example, apportioned the particulate matter for two cohorts living in urban and suburban Beijing (Shang et al., 2019).

Ferro noted that low-cost monitors should prove invaluable for determining what particles come from outdoors and which originate indoors. Mass balance models are now used regularly for assessing indoor $PM_{2.5}$ concentrations and human exposure to them.

Indoor Exposure Disparities

A 2011 framework developed by investigators at the Harvard T.H. Chan School of Public Health aimed to identify contributors to indoor exposure disparities (Figure 3-6) (Adamkiewicz et al., 2011). This framework acknowledges that there are disparities in the types of sources and their emissions depending on the specific community. Exposure factors, such as the built environment, number of people living in a residence, HVAC systems, air infiltration, and activity patterns, also vary by community. Ferro noted that looking at each of these factors systemically is important for addressing disparities, as is including the communities directly in this work.

Ferro raised the question of whether ambient air quality standards can be used to assess indoor air pollutants. In her opinion, the answer is no given that only some ambient particulate matter makes it indoors. As a result,

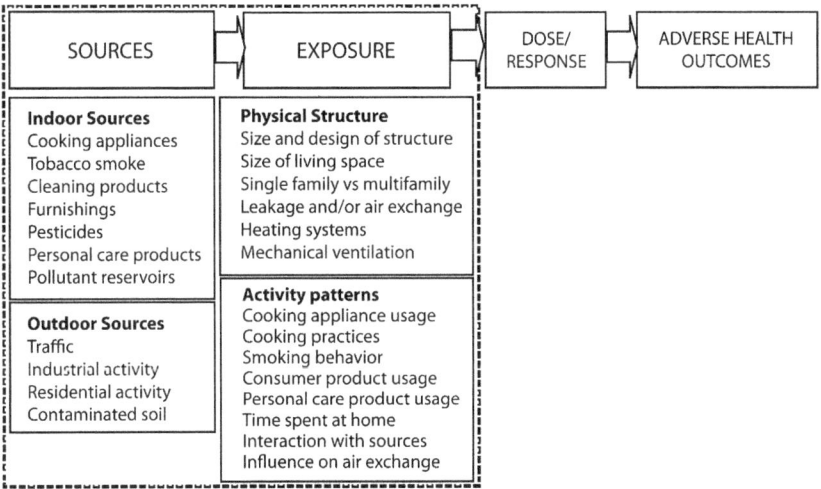

FIGURE 3-6 Framework for determining contributors to indoor particulate matter exposure disparities.
SOURCE: Ferro slide 15, from Figure 1 in the cited publication.[1]

[1] Adamkiewicz G, Zota AR, Fabian MP, Chahine T, Julien R, Spengler JD, Levy JI. Moving environmental justice indoors: Understanding structural influences on residential exposure patterns in low-income communities. *American Journal of Public Health*, 101(S1):S238–45, 2011. Reprinted with permission from the *American Journal of Public Health*.

applying ambient health-based standards to assess indoor $PM_{2.5}$ would underestimate risk. "We really need to separate the indoor-generated versus the outdoor-infiltrated particulate matter," said Ferro. One approach might be to use disability-adjusted life years as a way to prioritize indoor sources for mitigation. She noted that the Global Burden of Disease study[3] evaluates ambient $PM_{2.5}$ and ozone pollution as well as household air pollution from use of solid fuels for cooking to assess the health risks associated with air pollution. She also discussed the modular mechanistic framework (Eichler et al., 2021) as useful for modeling semivolatile organic compounds.

Ferro called for research to separate indoor-generated from outdoor-infiltrated $PM_{2.5}$ for epidemiological studies. She noted that it's also important to update human activity pattern data—for example, by using new technology like GPS-enabled devices—rather than relying on data from 30 years ago. In addition, advanced metrics and approaches are needed to assess and communicate the health effects of indoor $PM_{2.5}$ and to increase community-engaged research to better understand disparities in $PM_{2.5}$ exposure and develop mitigation strategies.

DISCUSSION

Planning committee chair Richard Corsi opened the discussion by asking the panelists to put the issues they discussed in the context of possible disparities in underserved and underresourced communities. Vance responded that there is a huge knowledge gap about what to do to reduce exposure in terms of indoor cooking emissions for those who are economically disadvantaged. The most effective approach to reducing such exposure is a good, externally venting hood or portable air cleaner, but many people cannot afford or are not able to take those steps, particularly if they are renting. A first step is to develop additional knowledge about human activity patterns, determine how personal habits differ with socioeconomic status, culture, and even region of the country, and use that information to inform better experiments to understand realistic exposures.

Waring said that for secondary organic aerosols, there are three questions that if answered could start to yield a picture of disparities and secondary aerosol exposure among those who live in underserved or poorer communities:

- Are there differences in outdoor concentrations for underserved communities or are they in regions of higher ozone and organic or inorganic particulate matter concentrations, which can infiltrate and potentially drive secondary chemistry?

---

[3] Information is available at http://ghdx.healthdata.org/gbd-2019.

- Are there differences in emissions of particular VOCs or precursors in housing types in underserved communities versus other communities, and are there different mixtures of VOCs present depending on the quality of the housing or the type of construction of the housing?
- Do individuals of different socioeconomic status use different consumer products that could generate more or fewer precursor chemicals?

Marr noted that she has been collaborating with a colleague on energy savings and building in underserved communities as these two factors pertain to indoor environmental quality. The residents of those communities, she said, are not concerned about ozone, volatile organic compounds, or plasticizers, but are very concerned about mold because they experience major asthma and allergy issues, which are more prominent in underserved communities. She wondered if the relatively sparse information in the literature on indoor exposure to mold has to do with the fact that this is predominantly a problem of underserved communities. Waring added that he was part of project in New York that measured mold in public housing, and it was everywhere. Elizabeth Matsui then noted that, while there are clear health effects associated with indoor $PM_{2.5}$ exposure, they are smaller than those from exposure to biological contributors such as pest allergens that are more common in underserved communities and endotoxin concentrations that correlate with mouse allergens in the indoor air.

Both Ferro and Prather commented on the need to include occupations in these studies. For example, there are investigations being done on exposures for workers in nail salons and those who clean office buildings, as well as other occupations subject to higher or different exposures.

Marr raised the problem of funding research in this area given that it tends to fall into an "interagency hole" where it is not clearly in the purview of the National Institutes of Health or the National Science Foundation. Farmer traced the problem to the fact that there are so many factors that may influence indoor air quality and health, making it difficult to assemble the bigger picture that would lead to mitigation and policies. "There is a clear need, and I think all of the talks highlighted that..., to not only think about this complex indoor air environment and what we can do to improve it, particularly in underserved communities, but also to start having cross-disciplinary discussions," said Farmer. Stephens added that resource constraints also limit what and who is studied, and Ivey stated that it boils down to money and the lower level of investment in the overall quality of life for people who are disproportionately exposed.

Planning committee member Seema Bhangar (WeWork) asked the panelists to discuss what they see as the biggest gaps in source control and if policies or regulations that address source control would directly address the most vulnerable populations. Ivey replied that the two biggest gaps concern the current regulations on fossil fuel extraction and combustion. "If you look at domestic oil and gas extraction, where the wells are, how far away they are allowed to drill from people's homes, there is a socioeconomic footprint and usually the burden is on communities of color." She noted that environmental groups are calling for methane and ethane to be removed from exemption from volatile organic chemical classification, which would allow for more strict emission standards for methane and ethane, which disproportionately affect disadvantaged communities. Marr added that she would like to see a standard or at least a benchmark for mold exposure.

Prather asked the panelists if enough is known at this point to create a framework for establishing which exposures are bad and which ones benign. Vance said that more information is needed about the sources of particulate matter of different sizes and the health effects associated with particular exposures to identify what should be of greatest concern. Waring wondered how basing those decisions on levels of indoor concentrations would work, given the lack of knowledge in this area. Farmer cited the importance of understanding the fundamental processes at work—the actual drivers of particulate matter composition and chemistry—to avoid controlling the wrong source.

Chan offered a final question, asking where resources should be spent in schools, given what is known about indoor particulate matter. Waring said that the greatest need in the Philadelphia schools he is familiar with would be adequate ventilation.

4

Day One Summary

To conclude the first day of the workshop, planning committee member Jeffrey Siegel (University of Toronto) recounted some of the themes that cut across several of the presentations. The first was that when considering indoor fine particulate matter, it is important to account for precursors and other things that can modify $PM_{2.5}$, such as the presence of vapor-phase material that can condense into particulates. He named five broad categories of influential factors: situational, building and HVAC, activity and mobility, occupant perceptions and behavior (noting that how people perceive the risks associated with particulate matter might modify their behavior, leading to changes in exposure and even how much is generated), and measurement. It was important to remember, he emphasized, that these factors cut across indoor and ambient particulate matter and were interrelated, not independent of one another.

Siegel noted a strong sentiment among the speakers for addressing the underlying causes of exposure disparities. This is, as Ivey observed, a systemic problem that requires systemic solutions, and the hunt for those solutions should be a primary driver of research needs.

Throughout the day, the speakers identified the need for research to

- Resolve exposure misclassification that arises from a lack of information about how the transport of ambient $PM_{2.5}$ to the indoors might change exposure.

- Develop health-relevant metrics and examine $PM_{2.5}$ exposure through the lens of what matters most to human health.
- Identify the chemical processes, deposition processes, and sinks that affect indoor particulate matter in general and $PM_{2.5}$ in particular.
- Measure unknown input parameters that go into models in order to apply the models more broadly.
- Characterize the composition of indoor $PM_{2.5}$ and the drivers that produce specific types of particulate matter.
- Explore the interactions between indoor $PM_{2.5}$ and the indoor microbiome.

Corsi closed the session by thanking the speakers and other participants for their contributions.

5

Health Effects of Exposure to Indoor Particulate Matter

The second day of the workshop focused on health, metrics, and assessment, and in his opening remarks Richard Corsi noted that there is overwhelming scientific evidence that increases in levels of outdoor fine particulate matter ($PM_{2.5}$) are associated with a range of short-term and chronic health effects, including asthma exacerbation, acute and chronic bronchitis, heart attack, increased susceptibility to respiratory infections, and premature death. Moreover, the burden of these health effects is greater for underserved and marginalized communities, something he asked the participants to keep in mind when listening to the presentations and discussions. He then posed three questions that the day's presentations would address:

1. What do we know about the health effects of exposure to indoor fine particulate matter?
2. How do we practically measure indoor fine particulate matter?
3. What do the measurements mean?

The day's first session featured three presentations by physicians on how exposure to indoor particulate matter can affect human health. Howard Kipen (Rutgers University School of Public Health) addressed the overall health burden associated with exposure to $PM_{2.5}$, and Meredith McCormack (Johns Hopkins School of Medicine) discussed the link between indoor $PM_{2.5}$ exposure and pulmonary disease, as well as disparities in economically challenged communities. Stephanie Holm (University of California, San Francisco) talked about the health effects of exposure to wildfire smoke and other ambient air pollutants and how certain building characteristics might mitigate those adverse health effects. Elizabeth Matsui and Linda McCauley (Emory University's Nell Hodgson Woodruff School of Nursing) comoderated an open discussion following the three presentations.

THE OVERALL (MOSTLY CARDIOVASCULAR) HEALTH BURDEN OF INDOOR $PM_{2.5}$ EXPOSURE

Howard Kipen outlined his presentation by saying that he would talk about what he perceived to be the confusion between the health effects of indoor and outdoor air pollution—a confusion that results from the fact that outdoor particulate matter penetrates indoors. He began by displaying data from the Global Burden of Disease study showing that exposure to household air pollution from solid fuels (which occurs primarily in less and least developed countries) and ambient particulate matter pollution are two of the 10 top causes of morbidity, expressed as disability-adjusted life years (Figure 5-1) (Lim et al., 2012). The biggest components of the morbidity associated with ambient particulate matter exposure are cardiovascular disability, chronic and acute respiratory disease, and infection, and these and other health effects (such as developmental issues, blood clotting, and metabolic problems) associated with particulate matter in general and $PM_{2.5}$ in particular are ubiquitous (Figure 5-2) (Thurston et al., 2017). He explained that because it is not always possible to study these clinical events, particularly when looking at their relationship to changes in air pollution over time, researchers tend to use biomarkers as proxies for both clinical events and exposures.

Taking an example from the PM_{10} literature (he asserted that $PM_{2.5}$ would be similar) Kipen showed the results of an analysis of administrative datasets generated from death certificates attributing mortality to respiratory disease, cardiovascular disease, and other causes. That analysis showed that the estimated percentage change in day-to-day death rates associated with each 10 microgram per cubic meter ($\mu g/m^3$) increase in PM_{10} is around 4 percent for respiratory disease, 2 percent for cardiovascular disease, and 1 percent for all causes of mortality. Kipen said that the takeaway message is that, while we know that the strength of the association between air pollution and respiratory disease is stronger, the vast burden of morbidity from air pollution is actually due to increased cardiovascular disease, largely because cardiovascular disease is so much more prevalent in adults living in the developed world.

It is likely that the mechanisms that link particulate matter and cardiovascular disease involve inflammation and oxidative stress that arise when particulate matter interacts with lung tissue and triggers physiological

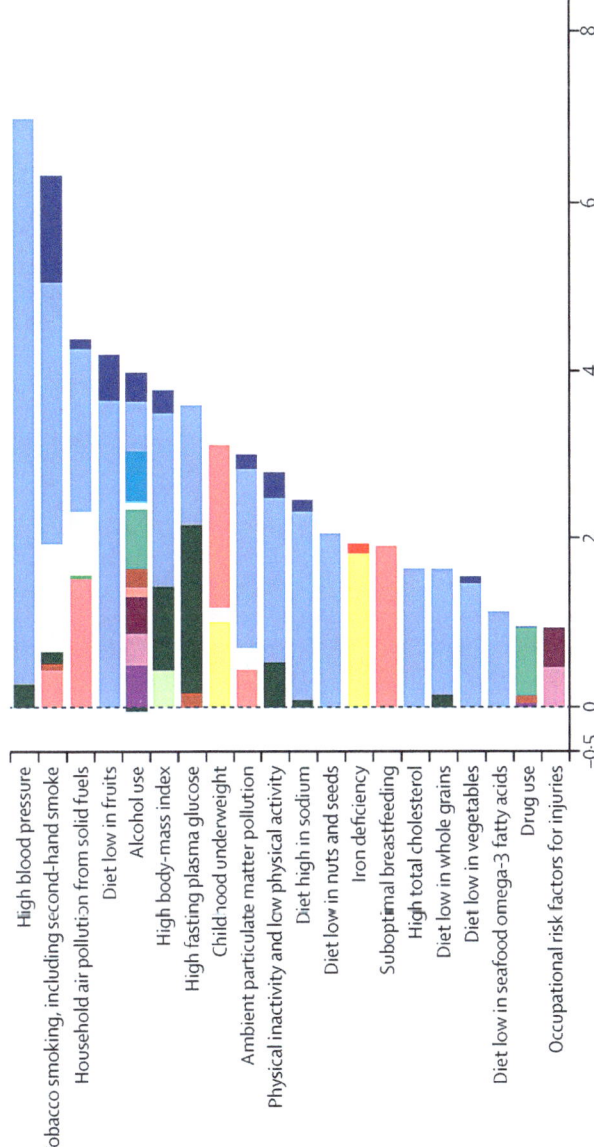

FIGURE 5-1 Global burden of disease attributable to 20 leading risk factors in 2010, expressed as a percentage of global disability-adjusted life years. Color coding for health outcomes available from the source.
SOURCE: Kipen slide 4, from Figure 2C in the cited publication.[1]
[1] Reprinted from *The Lancet* 380(9859) Lim SS, Vos T, Flaxman AD, Danaei G, Shibuya K, Adair-Rohani H, Amann M, Anderson HR, Andrews KG, Aryee M, and 202 others. 2012. A comparative risk assessment of burden of disease and injury attributable to 67 risk factors and risk factor clusters in 21 regions, 1990-2010: A systematic analysis for the Global Burden of Disease study 2010. pp. 2224–60. Copyright (2012), with permission from Elsevier.

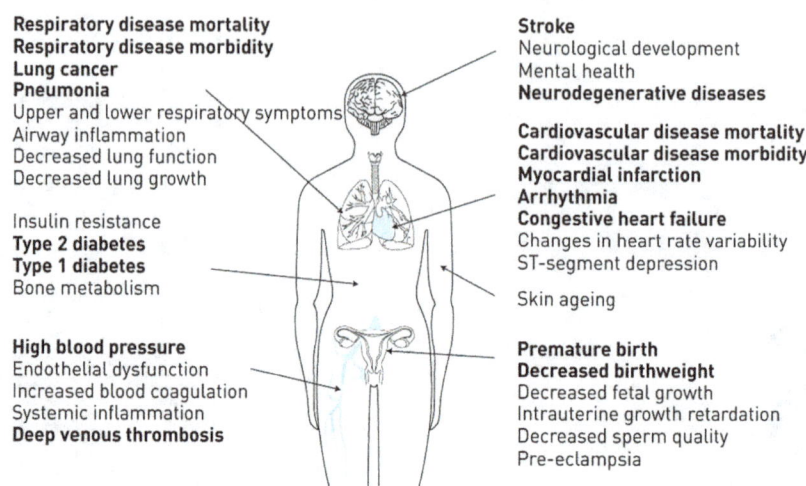

FIGURE 5-2 Overview of diseases, conditions, and biomarkers affected by outdoor air pollution. Bold type indicates conditions included in the Global Burden of Disease categories at the time of publication.
SOURCE: Kipen slide 5, from Thurston et al. (2017) Figure 1.[1]

[1] Reproduced with permission of the © ERS 2021: *European Respiratory Journal* 49 (1) 1600419; DOI: 10.1183/13993003.00419-2016 Published 11 January 2017.

events that can exacerbate atherosclerosis and, over time, lead to a heart attack (Figure 5-3) (Brook, 2008). Ultrafine particles that cross from the alveoli into the blood stream can also cause blood vessels to constrict and increase blood pressure.

Before the 2008 Beijing Olympics, Kipen participated in a study called the Health Effects of Air Pollution Reduction Trial. This trial involving healthy 20-year-olds living in Beijing tested the hypothesis that there would be biomarkers of systemic inflammation, endothelial dysfunction, increased coagulation, and autonomic dysfunction, as well as direct measures of oxidative stress, that would improve significantly during the period in which the Chinese government promised to reduce air pollution and that would then return to baseline in the post-Olympic period (Huang et al., 2012; Rich et al., 2012; Zhang et al., 2013).

During the "clean air" period, levels of all the pollutants measured, including $PM_{2.5}$, dropped significantly except for ozone, which naturally increases when nitric oxide levels fall. At the same time, biomarkers of pulmonary inflammation and oxidative stress, systemic inflammation and oxidative stress, asthma, coagulation, and autonomic tone, which affects heart rate and blood pressure, all improved. Unfortunately, when pollution levels increased once the Olympics ended, the biomarkers returned to where they

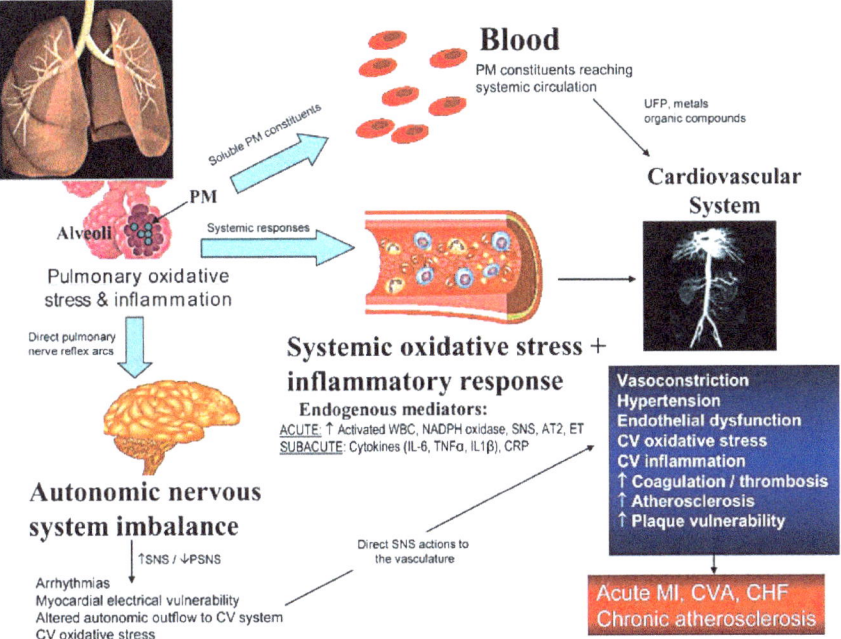

FIGURE 5-3 Possible mechanistic paths linking particulate matter (PM) exposure and cardiovascular (CV) disease. CHF = congestive heart failure; CVA = cerebrovascular accident; MI = myocardial infarction; PSNS = parasympathetic nervous system; SNS = sympathetic nervous system; UFP = ultrafine particles.
SOURCE: Kipen slide 9, from Figure 2 in the cited publication.[1]
[1] Republished with permission of Portland Press Ltd., from Brook RD. 2008. Cardiovascular effects of air pollution. *Clinical Science (London)* 115(6):175–87; permission conveyed through Copyright Clearance Center, Inc.

were before the clean air period. Kipen noted that this study did not measure heart attacks, atherosclerosis, or other clinical disease, but rather the biochemistry and cell biology that underlie those chronic diseases. "Most importantly, we saw that all of these changes were reversible in healthy young people," said Kipen. "So all of the health outcomes associated with exposure to $PM_{2.5}$, whether indoors or outdoors, might be reversible."

In another study, Kipen and his collaborators used a device that measures pressure inside the heart as a marker of heart failure to see if there was physiologic change associated with a 10 μg/m^3 change in ambient particulate matter. On days when there were increases in hospital admissions related to heart failure associated with increased ambient particulate matter, there was a corresponding increase in the pressure in the right ventricle of the heart. In essence, this work identified a biomarker for heart failure (Rich et al., 2008; Wellenius et al., 2006).

Epidemiology of Exposure to Indoor Air

Switching gears, Kipen said that the epidemiology of indoor exposure to air pollutants—including $PM_{2.5}$ and other constituents—has been done but isn't generally acknowledged because people are largely exposed to "outdoor" pollution when they are indoors, as the speakers in the workshop's first session described. "It turns out that indoor exposures to $PM_{2.5}$ probably account for 40 to 60 percent of total mortality attributed to $PM_{2.5}$ because even though we are measuring the pollution outdoors, that is not where we are," said Kipen. Based on that idea, he calculated that 71 percent of the average adult's dose of $PM_{2.5}$ comes from particles that were outdoors and infiltrated the indoor space.

After reviewing 16 studies that used an indoor air cleaner to lower household levels of $PM_{2.5}$, Kipen found that the air cleaners lowered indoor $PM_{2.5}$ levels between 40 percent and 82 percent but did not eliminate them. These studies also found that reduced indoor $PM_{2.5}$ levels correlated with reductions in systolic blood pressure and increases in peak blood flow, as well as nonsignificant reductions in endothelial function and inflammatory markers. While the potential for this type of research exists, one problem is that different researchers report tracking different biomarkers, making comparisons and meta-analyses less useful than they could be, he noted.

Kipen then discussed a study from 2005 in which his team exposed (with Institutional Review Board approval) 130 healthy, nonsmoking women in a test chamber to a mixture of VOCs and ozone similar to what might be found in so-called sick buildings (Laumbach et al., 2005). While the number of ultrafine particles rose significantly and then fell as they coalesced and formed $PM_{2.5}$ over time, there were no significant differences in nasal or upper airway symptoms or markers of nasal inflammation. Kipen concluded that VOCs and their oxidation products may not cause acute nasal effects at low concentrations, a finding similar to one in which laboratory rats were exposed to limonene and ozone (Sunil et al., 2007). The animal study did find some subclinical inflammatory changes in the lung that might be important in long-term exposure to indoor $PM_{2.5}$.

A more recent study found that household air cleaners can decrease the oxidative potential of indoor $PM_{2.5}$ when the data were normalized by volume, but not by mass (Brehmer et al., 2020). However, personal air monitor data failed to show any improvement in oxidative potential associated with air cleaning. Kipen noted that most of these studies have involved small numbers of individuals, which could explain the variable results.

On a final note, Kipen mentioned that he and his collaborators were now examining SARS-CoV-2 and how the virus may be associated with particles inside homes.

PULMONARY DISEASE ASSOCIATED WITH FINE PARTICULATE MATTER EXPOSURE IN INDOOR ENVIRONMENTS AND DISPARITIES IN ECONOMICALLY CHALLENGED COMMUNITIES

One of the unique characteristics of $PM_{2.5}$, said Meredith McCormack, is its ability to penetrate deep into the lungs and reach all the way to the alveoli, and the smallest components of $PM_{2.5}$, the ultrafine particles, can even translocate into the blood stream. As Kipen noted, it is possible to extrapolate from what is known about outdoor air quality to draw some inferences about the respiratory health effects of indoor $PM_{2.5}$, particularly on lung development, said McCormack.

The human lung continues to develop from birth until the early 20s, and during that time the alveoli increase exponentially and are susceptible to the health effects of air pollution, as has been shown with research on outdoor air exposures. In fact, studies have demonstrated that exposure to $PM_{2.5}$ affects the trajectory of lung development (Horak et al., 2002; Lee et al., 2010) and is associated with an accelerated decline in lung function with age (Doiron et al., 2019; Guo et al., 2018), McCormack explained. Conversely, improvements in air quality can improve optimal lung growth and attenuate the decline associated with normal aging.

Lung function is known to affect the risk of developing lung disease, and low lung function during early life is a risk factor for developing asthma during childhood and chronic obstructive pulmonary disease (COPD) later in life. In a study McCormack and her collaborators conducted in Baltimore City—where many people live in row houses with shared walls and in close proximity to a road—results showed that indoor $PM_{2.5}$ levels in children's bedrooms[1] were higher than outdoor levels and exceeded the annual outdoor limit for exposure set by the Environmental Protection Agency in three-quarters of the homes studied (McCormack et al., 2009). When compared to the results obtained from suburban homes, it was clear that inner-city children were exposed to much higher levels of indoor $PM_{2.5}$. Moreover, there was a consistently positive association between elevated $PM_{2.5}$ levels and asthma, other respiratory issues, and the need for children to use rescue medication (Figure 5-4).

In addition to spending time at home, children spend a good portion of their days in school environments—6–10 hours each school day. Few studies, however, have directly assessed school environmental conditions, in part because of the complexity of indoor air quality studies, said McCormack. She and her collaborators in Baltimore (Majd et al., 2019), as well as another group in Boston (Phipatanakul et al., 2011), have studied

---

[1] In-home levels were measured in the subject child's bedroom because that was the location where the child was expected to spend a substantial portion of their time indoors.

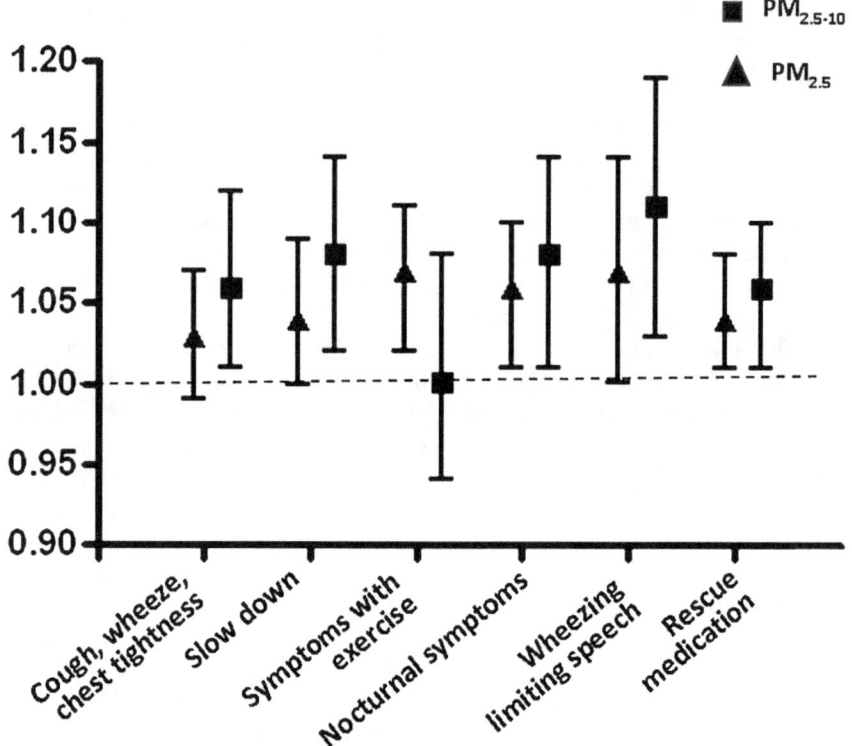

FIGURE 5-4 Indoor particulate pollution and asthma morbidity; coarse module adjusted for age, sex, race, parent education level, season, indoor fine PM, ambient fine PM, ambient coarse PM; fine module adjusted for age, sex, race, parent education level, season, indoor coarse PM, ambient coarse PM, ambient fine PM. IRR = incidence rate ratio; PM = particulate matter.
SOURCE: McCormack slide 8, adapted from Figure 3 in the cited publication.[1]

[1] Copyright © 2021 American Thoracic Society. All rights reserved. Breysse, P. N., Diette, G. B., Matsui, E. C., Butz, A. M., Hansel, N. N., & McCormack, M. C. (2010). Indoor air pollution and asthma in children. *Proceedings of the American Thoracic Society*, 7(2), 102–106. https://doi.org/10.1513/pats.200908-083RM. *Proceedings of the American Thoracic Society** is an official journal of the American Thoracic Society. Readers are encouraged to read the entire article for the correct context. The authors, editors, and The American Thoracic Society are not responsible for errors or omissions in adaptations.

*Now titled Annals of the American Thoracic Society.

air quality in inner-city schools. Both groups, said McCormack, found that the average air quality was within what she termed "reasonable limits" in the 30 schools they each studied, with an average $PM_{2.5}$ level of about 5 µg/m³. The Boston study, she added, found that outdoor factors contributed to indoor air quality, as did activities in the indoor spaces.

In Baltimore, about half the schools studied subsequently underwent a major renovation, and McCormack's team found that improvement in school facilities was associated with improvement in air quality and reduction in $PM_{2.5}$ concentrations. "So even though the baseline level was within a reasonable range, that modernization of school facilities led to even better air quality," said McCormack.

Indoor Air Pollution and COPD

Poor indoor air quality is a recognized contributor to COPD, particularly in low- and middle-income countries where indoor cooking using biomass fuels produces elevated levels of indoor $PM_{2.5}$. McCormack noted, though, that in many parts of the United States solid fuels (wood, coal, coke) are the primary heating source, accounting for more than 2.5 million households and 6.5 million people overall (Figure 5-5) (Rogalsky et al., 2014). Some 500,000 to 600,000 people who use solid fuels as their primary heating source live in households below the federal poverty line.

McCormack noted that several studies have demonstrated a relationship between solid fuel use for heating or cooking and COPD, as well as respiratory disease, even among people who have never smoked tobacco (Barry et al., 2010; Raju et al., 2019a,b). Other studies have found that even in homes where $PM_{2.5}$ levels were fairly low, around 11 µg/m³, indoor $PM_{2.5}$ was associated with increases in severe COPD exacerbations, respiratory symptoms, and rescue medication use among former smokers (Hansel et al., 2013). During warm months, there was a synergistic effect of indoor heat and $PM_{2.5}$ exposure that increased COPD exacerbations, respiratory symptoms, and rescue inhaler use (McCormack et al., 2016).

Susceptibility Factors and Interventions

To try to understand whether there are distinct individual factors that might increase susceptibility to the adverse health effects of $PM_{2.5}$, McCormack collaborated in a study led by her colleague Elizabeth Matsui in which they looked at the association between asthma-related symptoms and weight in 150 children. The results showed that overweight and obese children had increased nocturnal and exercise-related symptoms and that there was a synergistic effect of body mass and $PM_{2.5}$ levels on asthma symptoms (Lu

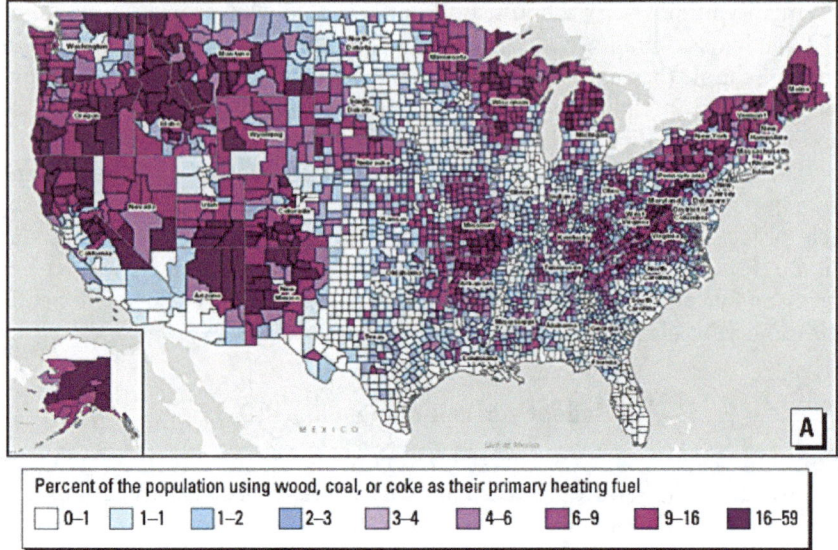

FIGURE 5-5 Solid fuel use as primary heating source (data from 2006–10).
SOURCE: McCormack slide 17, from Rogalsky et al. (2014) Figure 1A; reproduced from *Environmental Health Perspectives* with permission from the authors.

et al., 2013). The investigators found a similar trend when they examined the effect of exposure to secondhand smoke on asthma symptoms.

More recently, McCormack and her collaborators were able to measure tidal volume[2] in children with asthma and found an association between increasing body mass and increasing tidal volume. Modeling results showed that body mass was associated with increased particulate matter deposition rate. "Potentially, this is a characteristic important among children with obesity," said McCormack, who added that she has seen the same association in adults with COPD (McCormack et al., 2015), suggesting that obese individuals have a greater adverse response to particulate pollution.

Interventions can reduce $PM_{2.5}$ levels. One study, for example, found that an air cleaner intervention in homes with an adult smoker reduced indoor $PM_{2.5}$ levels by 50 percent and improved symptoms and increased the number of symptom-free days by 14 to 18 percent in children with asthma (Butz et al., 2011). McCormack noted that a similar effect size was seen in a trial with adult former smokers who have COPD (Sampson and Holgate, 1998).

---

[2] Tidal volume is the amount of air that moves in or out of the lungs with each respiratory cycle.

In conclusion, McCormack noted four primary implications from the studies she spoke about:

1. Indoor particulate pollution is a contributor to health disparities.
2. Indoor particulate pollution is associated with increased asthma and COPD morbidity.
3. Obesity may increase susceptibility to particulate pollution health effects.
4. Improving indoor air quality represents a therapeutic target to modify disease activity.

WILDFIRE SMOKE AND OTHER AMBIENT AIR POLLUTION COME INDOORS: HEALTH EFFECTS AND THE BUILDING CHARACTERISTICS THAT MITIGATE THEM

Wildfire events can expose large populations to smoke, over regions that extend far beyond the fire, said Stephanie Holm. "Even if you are not located close to a wildfire smoke event, you can still be affected by smoke from that event," said Holm. She noted that, while seeing smoke is a cause for concern, visibility alone isn't a reliable indicator of potential risk because conditions such as fog and rain can affect it. "So just because you can't see smoke, if the air quality index is bad you should be concerned," she explained. She added that, though a common perception is that California has the biggest concerns about wildfires, the National Oceanographic and Atmospheric Administration estimates that the risk of very large fires in the mid-21st century, compared to the end of the 20th century, extends to large portions of the United States (Figure 5-6).

As Kipen and others had noted throughout the workshop sessions, roughly half of the outdoor $PM_{2.5}$ comes indoors (Azimi and Stephens, 2020). Moreover, in terms of an individual's total exposure to $PM_{2.5}$, roughly 53 percent is outdoor $PM_{2.5}$ that people are exposed to in their home or other buildings. Holm observed that many air pollution epidemiologic studies use residential location as a proxy for exposure to outdoor ambient air pollution, although some of that exposure is indoors. For example, a 2017 Medicare study with 460 million person-years of follow-up found that a 10 µg/m^3 increase in $PM_{2.5}$ was associated with an approximately 7 percent increase in mortality (Di et al., 2017). The Children's Health Study in Los Angeles, which followed three cohorts involving more than 2000 adolescents, found that a 12.6 µg/m^3 decrease in $PM_{2.5}$ was associated with a significant increase in lung function (Gauderman et al., 2015). Conversely, the Imaging Dementia—Evidence for Amyloid Scanning (IDEAS) Study found that a 10 µg/m^3 increase in $PM_{2.5}$ increased the odds

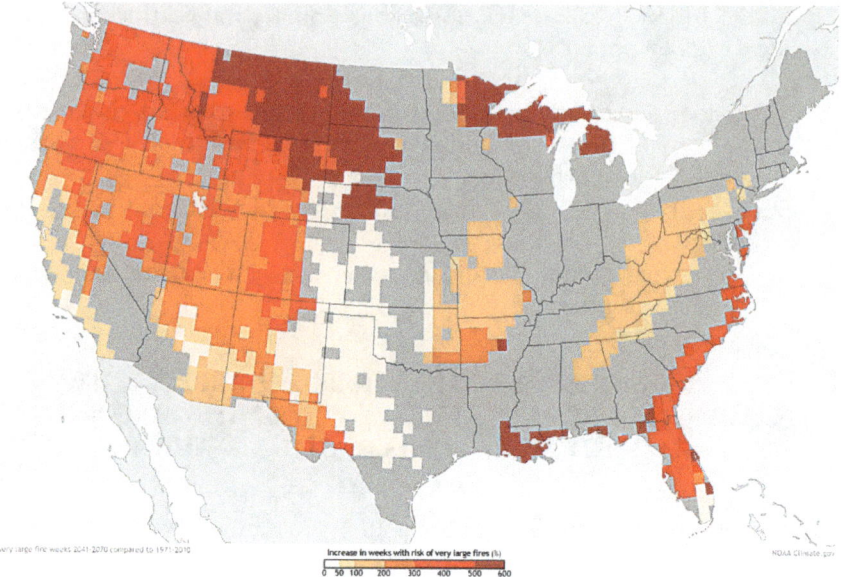

FIGURE 5-6 US regional estimates for weeks with a risk of very large fires in the mid-21st century compared to the end of the 20th century.
SOURCE: Holm slide 5, from Kennedy (2021).

of finding amyloid (which is thought to play a role in Alzheimer's disease) during a positron emission tomography brain scan (Iaccarino et al., 2021).

After reviewing how infiltration, mechanical ventilation, and natural ventilation determine how much outside $PM_{2.5}$ comes inside, Holm noted that while it might be tempting to think that a solution is to close up these three avenues into a home, doing so could lead to exposure problems from indoor-generated $PM_{2.5}$ (as well as moisture and mold). The key to addressing this problem is filtration using a filter with a minimum efficiency reporting value (MERV) rating of 13 or higher for central heating and air conditioning systems. For portable air cleaning systems, the unit needs a high enough clean air delivery rate to account for the size of the home and should not produce ions or ozone.

Interventions and Health Effects

Studies have shown that weatherization that decreases infiltration, combined with ventilation and education on intervention strategies, can improve asthma symptoms and overall self-reported health (Breysse et al., 2011, 2014). In contrast, decreasing infiltration without improving ventilation or filtration can actually increase indoor $PM_{2.5}$ exposure (Yuan et al., 2018) and might increase the rate of severe asthma events (Fabian et al., 2014). In schools, improved ventilation has been associated with fewer absences (Mendell et al., 2013) and improvements in math and language performance (Wargocki and Wyon, 2007). "When thinking about improving exposure to $PM_{2.5}$ from outdoor sources, ventilation needs to be accompanied by filtration," Holm added, but she cautioned that ion generation technologies may negate the benefits of filtration (Allen and Barn, 2020) and suggested sticking to mechanical air filtration.

Several studies have shown that improved filtration yields health benefits. For example, it is associated with an increase in birth weight among full-term infants (Barn et al., 2018). In children, improved filtration has been associated with a decrease in asthma morbidity (Martenies and Batterman, 2018), but does not decrease airborne nicotine exposure if there are smokers in the house, which may explain why filtration has fewer benefits in such households (Gold et al., 2017).

A few studies have looked at interventions to reduce exposure to wildfire-generated $PM_{2.5}$. A study in Australia, for example, showed that tightening the building envelope reduced peak $PM_{2.5}$ levels by as much as 76 percent (Reisen et al., 2019), and a study in Southern California found that running a home's heating and air conditioning system produced marked reductions in indoor wildfire-generated $PM_{2.5}$, particularly when running continuously with filters rated MERV-9 or higher (Fisk and Chan, 2017). Studies have also found portable air cleaners to produce substantial decreases in indoor $PM_{2.5}$ during wildfire events (Barn et al., 2008; Henderson et al., 2005; Stauffer et al., 2020). Holm noted, however, that total $PM_{2.5}$ infiltration may be higher during wildfire smoke events as a result of filters getting saturated and because there may be more ultrafine particulate matter that filtration captures less efficiently (Mendoza et al., 2021).

With regard to the effect of interventions on health outcomes during wildfire events, a modeling study found that MERV-13 filtration and a portable air cleaner installed in every home in Southern California during the 2003 wildfire season could have prevented more than half of the hospital admissions due to respiratory issues (Fisk and Chan, 2017). Another study found that filtration was the only intervention that benefited health among residents of the Hoopa Valley National Indian Reservation in northwestern California (Mott et al., 2002); mask wearing and leaving the area did not confer the same health benefits.

Holm closed by noting that the Western States Pediatric Environmental Health Specialty Unit, where she is codirector, has produced a number of educational resources on reducing wildfire smoke exposures for children.[3]

DISCUSSION

Matsui opened the discussion by asking the speakers to identify important research gaps that need to be filled. Kipen replied that because biomarkers may not always be suitable for studying health effects, there is a need to identify markers related to "real" endpoints that have never been measured. He noted that the National Institute of Environmental Health Sciences may have some requests for application coming soon in this area. McCormack noted that indoor air studies can be challenging because of the intensity of the study protocol, which leads to smaller sample sizes that make it difficult to identify multisystem effects that may be downstream from pollution exposure. She would like to see studies that look at more than one outcome in at-risk populations and longitudinal studies that measure downstream consequences of exposure to particulate matter. Holm agreed with all of these suggestions and added that she is concerned that there is not sufficient characterization of the difference in exposure profiles for indoor versus outdoor particulate matter and the effects on health in the broader population. She noted that there are gaps in understanding the effects of short-term spikes in $PM_{2.5}$ exposure as opposed to levels that are averaged out over time. The recent availability of low-cost exposure monitoring technologies may help.

Comoderator Linda McCauley, impressed with the data on intrusion of outdoor particulate matter into the indoor environment, said that she would like to see public health messages reflecting such data. She noted that the vast majority of public health messages today concern when it is safe for outdoor play and recreation. Many people, however, do not believe they have an indoor air quality issue. Holm agreed with McCauley, noting that the natural response in schools is to keep children inside when the outdoor air quality is poor without considering the school's indoor air quality. She reported that Washington State is looking at what the messaging should be regarding indoor and outdoor levels, and that the EPA and several Pediatric Environmental Health Specialty Units[4] are collaborating on a series of

---

[3] These resources are available at https://wspehsu.ucsf.edu/projects/wildfires-and-childrens-health-2/.

[4] The Pediatric Environmental Health Specialty Units are a national network of experts in the prevention, diagnosis, management, and treatment of health issues that arise from environmental exposures from preconception through adolescence. Additional information is available at https://www.pehsu.net/.

workshops on how to message during wildfire smoke events considering indoor versus outdoor exposures.

McCormack added that the EPA's Tools for Schools[5] and its anti-bus-idling campaign are good examples of a focus on the contribution of outdoor sources to indoor $PM_{2.5}$ levels and making actionable recommendations. She also suggested that the COVID-19 pandemic has provided a unique opportunity to educate the public about the importance of indoor air quality. Kipen observed that an additional consideration was the possibility that residences that were tightly sealed to prevent infiltration could be subject to increases in exposure to other indoor contaminants, which may result in their own health problems.

Responding to a question about ultrafine particulate matter, McCormack noted that new technologies are now making it possible to study levels and health effects of ultrafine particulate matter and enhancing the ability of air cleaners to reduce exposure to it.

Jeff Siegel noted that studies on interventions have not focused on how variations in their performance—for example, in how an air cleaner performs under different conditions or over time—affect the data being generated. McCormack agreed that this is a challenge, and Kipen added that indoor studies have also been too short to truly judge the effectiveness of the interventions being tested.

Marina Vance asked the panelists for their thoughts on what a framework to assess the health effects from indoor sources of particulate matter would look like and if there would ever be a place for a guideline for indoor $PM_{2.5}$. Holm endorsed such a guideline, and maybe even a standard; from her perspective of working with schools, she finds it hard to advise them on steps to take because there is no guideline or standard for indoor $PM_{2.5}$ exposure. McCormack agreed and supported conducting the research needed to develop a standard and put one in place. Schools, she said, might be a good place to start that work. McCauley followed up by asking if systematic monitoring of indoor levels might be a part of such an effort and McCormack replied that that would be useful and that newer technologies made it possible.

Both McCormack and Holm noted there is an environmental justice component to this, given that public health officials during a smoke event may tell schools to send their students home, but that assumes that the indoor air quality at home will be better than at the school. "That is especially unlikely to be true for our lower-resourced families," said Holm, "so we are exacerbating preexisting disparities for those children." In her opinion, the way to improve access to good-quality indoor air for all children is to ensure that schools have clean indoor air.

---

[5] https://www.epa.gov/iaq-schools

6

Indoor Exposure to Particulate Matter: Metrics and Assessment

The second session of the day featured three speakers. William W. Nazaroff (University of California, Berkeley) discussed indoor $PM_{2.5}$ measurement, exposure, and control. Kirsten Koehler (Johns Hopkins Bloomberg School of Public Health) then addressed the challenge of moving from measuring indoor $PM_{2.5}$ to evaluating a building occupant's exposure. Dusan Licina (Swiss Federal Institute of Technology, Lausanne) spoke about low-cost indoor particulate matter sensors and how they can be used and misused. Following the three presentations, Elizabeth Matsui moderated an open discussion.

TRANSCENDING COMPLEXITY: INDOOR $PM_{2.5}$ MEASUREMENT, EXPOSURE, AND CONTROL

William Nazaroff opened by noting that an estimated 6 million of the 55 million deaths worldwide in 2016—more than 10 percent—were associated with air pollution, and 4.1 million were tied to ambient particulate matter pollution. Secondhand tobacco smoke, a major contributor to ill health, was linked to nearly 1 million deaths, primarily because of its $PM_{2.5}$ content. For context, he noted that some 2.9 million people worldwide had died of COVID-19 as of April 9, 2021. Again, the majority of deaths associated with exposure to $PM_{2.5}$ are most likely caused by indoor exposure and most likely by residential exposure, given the percentage of time that people typically spend in that environment.

For a time, Nazaroff worked in Singapore, where there is seasonal wildfire smoke from agricultural burning, and during one of these smoke episodes several of his collaborators carried personal particulate matter monitors with them as they went about their daily routines. Figure 6-1 shows one day's results for someone who spent the night in a hotel, walked to a building called the CREATE (Campus for Research Excellence and Technological Enterprise) Tower, spent time in the tower, took a break at lunch and walked outside for a time, and then after another period of work in the tower walked back to the hotel and spent the night (Zhou et al., 2015). Overall, the hotel and the research tower afforded about a 50 percent reduction in the indoor concentration of $PM_{2.5}$.

While indoor sources of $PM_{2.5}$ were not particularly important in this case, indoor emission sources such as smoking, burning candles, cooking, breathing, humidification, and vacuum cleaning can be important contributors in the home. In Nazaroff's view as an engineer, there are four dimensions of complexity when thinking about indoor $PM_{2.5}$: size, temporal, chemical, and spatial complexity. For the bulk of the $PM_{2.5}$ mass, particles range in size from 0.01 to 2.5 microns, a 250-fold span in diameter and a 16 millionfold span in mass between the largest and smallest particles, which Nazaroff pointed out is the same mass ratio between an Etruscan shrew and a gray whale.

Chemically, the composition of fine particles incudes elemental and organic carbon, with vastly diverse chemical composition, as well as crustal materials, inorganic salts, metals, microbes, and other constituents. Temporal complexity arises from the fact that indoor concentrations can change

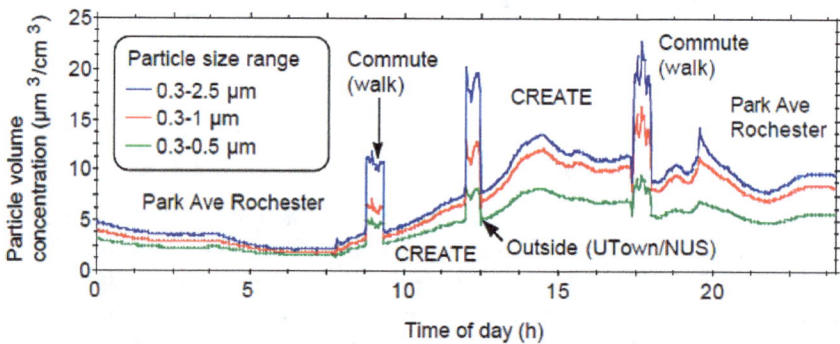

FIGURE 6-1 One person's $PM_{2.5}$ exposure in Singapore on June 25, 2013, during a wildfire smoke event. CREATE = Campus for Research Excellence and Technological Enterprise; UTown/NUS = University Town/National University of Singapore.
SOURCE: Nazaroff slide 4, adapted from Figure 2 in the cited publication.[1]

[1] Reprinted from *Building and Environment*, vol. 93, Zhou J, Chen A, Cao Q, Yang B, Chang VWC, Nazaroff WW. 2015. Particle exposure during the 2013 haze in Singapore: Importance of the built environment, pp. 14–23, Copyright (2015), with permission from Elsevier.

by a factor of 100 in a matter of minutes, in response, for example, to an episodic emission event such as cooking. Spatially, each building is its own distinct entity with its own fine indoor particulate matter level, and even within a building concentration may vary among rooms and within a room.

Size and spatial complexity exist, too, in the distribution of different size particles in the three major regions of the respiratory tract (Figure 6-2) (Yeh et al., 1996). Nazaroff explained that this differential distribution of particles has consequences for the body's physiologic response, as there are different clearance mechanisms and different disease sensitivities in different regions of the lung.

Chemical complexity occurs in multiple forms. For example, when outdoor $PM_{2.5}$ comes indoors, its chemical composition changes as the temperature and humidity change and those changes alter the equilibrium partitioning for semivolatile species, including water. Particle surfaces take up and emit gaseous species indoors, which can affect the condensation to and evaporation from the particles, and there can be shifts in aerosol liquid water and pH, which can alter the size and composition of the particles (Goldstein et al., 2021).

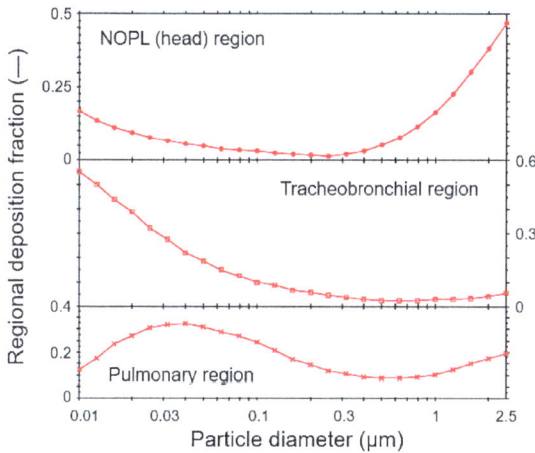

FIGURE 6-2 Size and spatial complexity of particle deposition in different regions of the lung. NOPL = nasal, oral, pharyngeal, and larynx.
SOURCE: Nazaroff slide 7, adapted from Figure 1 in the cited publication.[1]

[1] Comparisons of Calculated Respiratory Tract Deposition of Particles Based on the Proposed NCRP Model and the New ICRP66 Model. Yeh H-C, Cuddihy RG, Phalen RF, Chang IY. *Aerosol Science and Technology* 25(2):134–40. ©The American Association for Aerosol Research; http://www.tandfonline.com on behalf of The American Association for Aerosol Research, reprinted by permission of Taylor & Francis Ltd.

Activity and Occupancy Produce Complexity

Concentrations vary markedly indoors, mostly in response to changing indoor conditions, especially indoor emission activities, said Nazaroff (Patel et al., 2020). He reminded listeners that the data Marina Vance presented from the HOMEChem study (see Chapter 3 for details) clearly showed the variations in indoor particulate matter concentrations throughout the day as different activities occur inside the home. He explained that what this highlights about temporal variability is that relying on time-averaged indoor concentration measurements without accounting for the correlation with occupancy can mischaracterize indoor exposure.

As an example, Nazaroff discussed the results of a study in which his research group outfitted a small university classroom with a particle sensor and measured particle distribution throughout the day as the classroom sat empty or was occupied (Figure 6-3) (Qian et al., 2012). The results showed that smaller particles, those 0.3 to 0.5 microns in diameter, were not influenced by the occupants, while the largest particles were strongly influenced

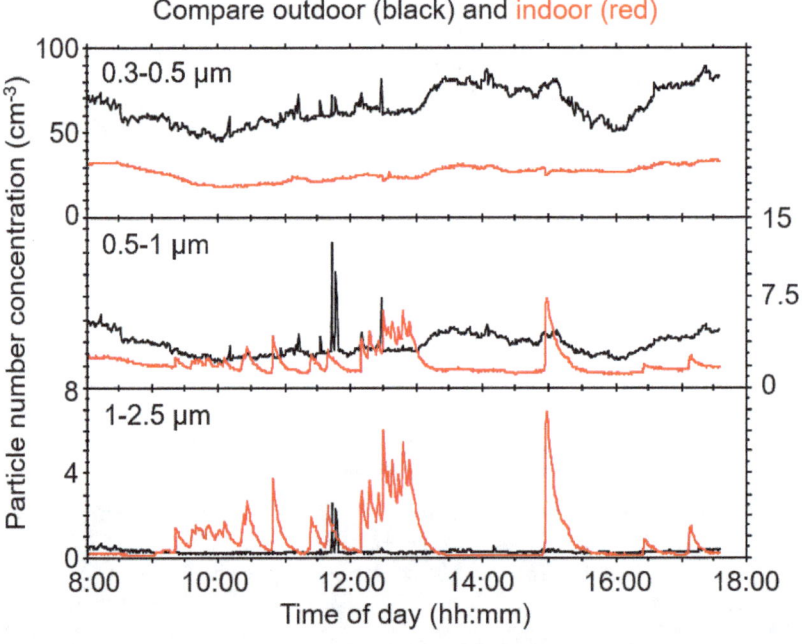

FIGURE 6-3 Temporal complexity's correlation with occupancy, using the example of optical particle counter data from a university classroom in normal use.
SOURCE: Nazaroff slide 10, from Qian et al. (2012), adapted from Supplementary Figure S7; reprinted with permission from John Wiley and Sons, Inc.

by occupants and movement in the classroom thanks to particle shedding from the occupants and resuspension of particles from the floor.

In another study using a well-controlled chamber to study spatial complexity, Nazaroff and his collaborators had subjects conduct different activities while the research team measured particle levels in their breathing zones and in the background room air (Figure 6-4). Levels in the breathing zone were considerably higher than the room average levels when the occupants were seated, but when they walked around the room the personal particulate matter cloud dissipated because the room floor was clean and the room air became relatively well mixed by their movement. He noted that the spatial variability between what is measured in the breathing zone and room background is referred to as the "Pigpen effect," after the Charles Schultz character in *Peanuts*.

Simplifying the Complexity

Nazaroff discussed several approaches for getting past the complexity. For size complexity, he suggested using $PM_{2.5}$ mass concentration or some suitable proxy as the source of best information if the aim is practicality. For ultrafine particles ($PM_{0.1}$), it is probably best to use number concentration, though this is an area of ongoing challenge, he said. For chemical

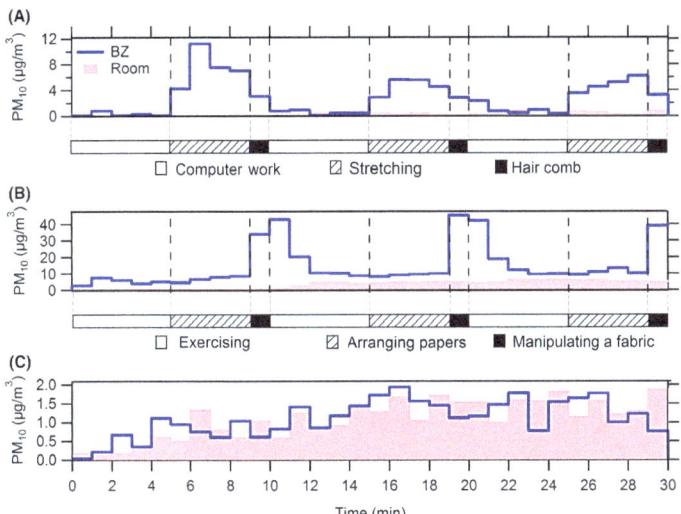

FIGURE 6-4 Measurements of particulate matter (PM) in a controlled chamber over time. BZ = breathing zone.
SOURCE: Nazaroff slide 11, from Licina et al. (2017) Figure 2; reprinted with permission from John Wiley and Sons, Inc.

complexity, he noted that the evidence to date does not implicate any compositional factors as a primary reason for adverse health consequences from particulate matter exposure. "I would say that from a practical orientation right now, if we aim to improve indoor environmental conditions, chemical complexity should be made secondary," he said, adding that this is an active research area and that perspective could change with more information.

Addressing temporal complexity requires being mindful that measurements effectively represent conditions during occupancy and not misinterpreting time-averaged data to represent what occurs when people are present. In terms of spatial complexity, Nazaroff explained that personal monitoring is better than stationary monitoring, and that indoor data are better than outdoor data.

He cautioned that no instrument is capable of measuring every aspect of particulate matter. The key considerations when choosing a particular measurement method for the indoor environment include the particle size range and size resolution to be studied, if time-resolved versus time-integrated sampling is needed, whether chemical composition information is desired, and device cost, portability, and performance stability. Among the alternative measurement methods are

- light scattering, which is good for particle sizes ranging from 0.3 to 10 microns
- electrical mobility to produce size-resolved data for ultrafine particles
- condensation particle counters for measuring ultrafine particle number concentrations
- filter-based sampling, a standard method for time-integrated mass concentration and some chemical composition data.

He noted that practical constraints are different for personal monitoring compared to indoor air sampling.

Twelve Words to Improve Indoor Air Quality

Nazaroff spent the final portion of his presentation on ways to close the gap between what we know about improving indoor air quality and what we should do to achieve improvement. This is different, he clarified, from the gap between what we ought to know and what we do know, which is the subject of ongoing research.

His guideposts for improving indoor air quality come down to four ideas encapsulated in 12 words:

1. minimize indoor emissions,
2. keep it dry,
3. ventilate well, and
4. protect against outdoor pollution.

Minimizing indoor emissions is not new guidance. The biggest improvement in indoor air quality that has occurred in his lifetime is the massive reduction in exposure to environmental tobacco smoke resulting from a combination of actions, including public messaging and regulation. Cooking is another target that could be addressed with a range hood and its effective use, and personal emissions of infectious agents can be controlled with masking. Keeping the environment dry can avoid mold and related bioaerosol problems.

Ventilation is a complex issue that is hard to simplify. If the air exchange rate is too low indoor pollution sources will dominate, but if it is too high it will take too much energy and outdoor pollution sources such as wildfire smoke will dominate. "We need to find the Goldilocks solution here, which is to ventilate sufficiently but not excessively," said Nazaroff, noting that research has yet to identify a solution.

For the final guidepost, protecting against outdoor pollution, filtration is one solution, and it can take two forms that Nazaroff called the closed and open paths. In a closed-path intervention, protecting occupants of an indoor space from outdoor pollution involves putting a filter in a mechanical supply air stream that would remove particles with some efficiency. If the intervention can avoid bypass and leakage, the goal would be to strive for the most efficient and practical device. This approach, however, does not protect occupants from indoor sources.

An open path intervention uses a recirculating air filter with a fan unit indoors, and efficiency is measured by the fraction of particles removed as the air passes through the filter. Nazaroff explained that it is possible that this control measure will both reduce exposure to outdoor originating particles and reduce the effect of emissions indoors on exposure, but it is an open path because both the outdoor particles and indoor particles can get to a susceptible individual without having passed through the filter. He also noted three possible performance metrics for indoor particulate matter filtration: efficiency, clean air delivery rate, and effectiveness. Efficiency is the fractional removal of particles, the clean air delivery rate combines efficiency with the flow rate through the device, and effectiveness refers to the extent to which the controlled condition is better than the uncontrolled condition.

When protecting against outdoor pollution, efficiency is the same as effectiveness for a closed-path intervention. The key to performance, then, is to make the fractional removal of particles as high as possible. With the recirculating systems used in an open-path intervention, efficiency and effectiveness are not the same, and the goal should be to achieve a high clean air delivery rate compared to the removal rates that occur by other processes, explained Nazaroff. An important point here, he added, is that effective particle filtration can trade off higher flow rate for somewhat lower efficiency to obtain good effectiveness.

In closing, Nazaroff cited a quote attributed to Albert Einstein: "Everything should be made as simple as possible, but not simpler." The intervention opportunities with the biggest potential to improve human health include a combination of resident and school environment improvements.

THE CHALLENGE OF MOVING FROM THE MEASUREMENT OF INDOOR PM$_{2.5}$ TO EVALUATING OCCUPANT EXPOSURE

Kirsten Koehler began by presenting an environmental health paradigm developed some 30 years ago (Figure 6-5) (Sexton et al., 1992). It shows the trajectory of a contaminant released from some source and transported or transformed in the ambient environment (air, water, and/or land). Once it comes in contact with an individual, there is a potential for exposure, which can occur at the individual, community, or population level. Inside the body, a contaminant can lead to a dose, and ultimately to altered structure and function, and adverse health outcomes. Koehler clarified that for particulate matter, "dose" may not be the best measure of exposure because the relationship between exposure and dose is a strong function of particle size. "Dose is not as easy for biomarker indicators for particulate matter…

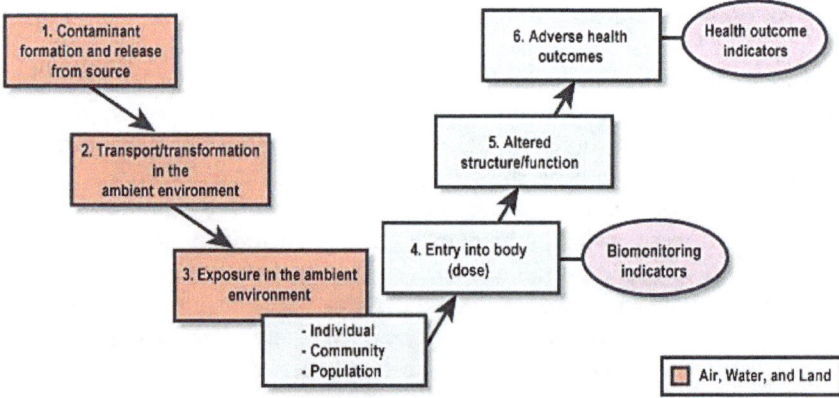

FIGURE 6-5 EPA environmental public health paradigm.
SOURCE: Koehler slide 2 (https://www.epa.gov/report-environment/human-exposure-and-health).

there is not some magic bullet we can measure in urine that tells us everything about particulate matter exposure," she said.

For particulate matter, it can be useful to add more steps to the paradigm, such as transformation in the environment that leads to some distribution of air pollution exposures across a city. A person in that city would breathe a fraction of those particles, which would deposit in different regions of the respiratory tract, producing a dose and ultimately a health effect, said Koehler. In the United States, regulations control the emissions from some sources and industries, but many factors can affect the relationship between emissions and ambient concentrations, even close to sources.

Exposure, which can have many meanings, often refers in this case to a so-called "ecological estimate" in which exposure estimates for groups of people are based on some common location, activity, or source. This conceptualization of exposure is most useful, said Koehler, when assigning exposure estimates to large groups of people. For $PM_{2.5}$ exposures, the data for these estimates might come from the widely distributed EPA Air Quality System database and be determined using information from the monitor nearest a group's neighborhood. A low-cost sensor network could enable a more detailed mapping of $PM_{2.5}$ concentrations and support interpolation across a city to the level of an individual's home, but few cities have a suitable density of monitors to achieve that degree of resolution for all their residents.

Such monitoring is important, Koehler noted, as there can be large-scale changes in particle concentration and composition even close to a source such as a major roadway (Zhu et al., 2002). "We expect that there are large spatial gradients and concentrations, and this provides an opportunity to think more about how low-cost sensor networks could help us understand this variability," said Koehler.

Personal Exposure Monitoring

Even a dense network of low-cost sensors would not yield information on all the important sources of indoor particulate matter or the various discrete exposures an individual experiences during the day. This is where personal monitoring comes into play. Koehler noted that by pairing personal monitor measurements with GPS data and air pollution measurements of high enough temporal resolution it would be possible to attribute exposures to the different microenvironments in which people spend time. There are a number of sensors on the market, she added, that can collect data at a per-second temporal resolution.

When conducting exposure assessments, it is important to consider whether an acute or chronic exposure is the more important factor in terms of health effects. Koehler said that the time between exposure and the onset of a health outcome is another important consideration, and, because it is unlikely that researchers will ask people to carry air quality monitoring equipment with them for an entire year, it may be necessary to take an ecological approach to assessing the link between exposures and health outcomes.

To illustrate the type of data that personal exposure monitors can generate, Koehler described the results of a study in which she and her collaborators asked 44 healthy adults in Fort Collins, Colorado, to carry personal monitors with them for 8 nonconsecutive days (Figure 6-6). The results show the large variability from the lowest to highest cumulative exposures as well as the variable proportions of exposures from different microenvironments (Koehler et al., 2019). Cumulative exposures, she explained, reflect the fact that people do not spend the same amount of time in each microenvironment and that one microenvironment can make a large contribution to total exposure even if the time spent in it is brief. When she and her colleagues conducted a similar study, in which 50 asthmatic children in Baltimore carried a backpack with air sampling equipment for 4 consecutive days, they found that home exposures were higher than ambient exposures for some of the children.

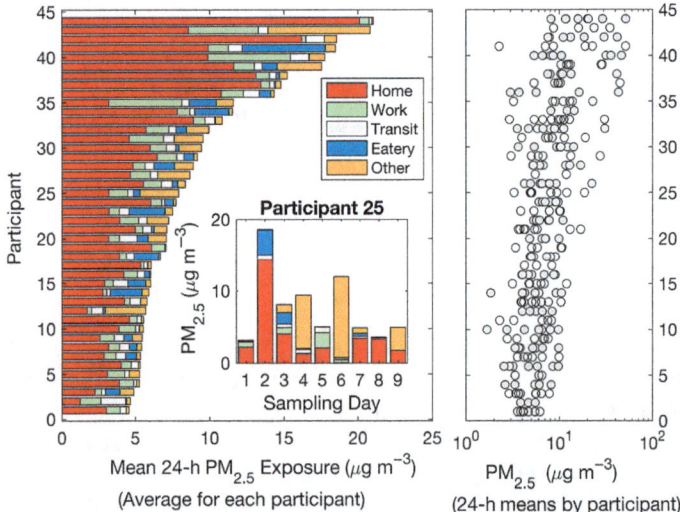

FIGURE 6-6 Variability of $PM_{2.5}$ exposures by person, day, and microenvironment. The panel on the right shows cumulative $PM_{2.5}$ exposure averages over 8 nonconsecutive days.
SOURCE: Koehler slide 13, from Koehler et al. (2019) Figure 3; reprinted with permission from John Wiley and Sons, Inc.

As other speakers noted, not all particle sizes penetrate to all regions of the respiratory tract. Close to 100 percent of particles smaller than 1 micron and 50 percent of 100 micron particles are able to penetrate into the respiratory tract. From there, only about half of 10 micron particles can make it into the thoracic region, and about half of the particles that are 4 microns or smaller can penetrate deep into the lungs. Koehler noted that not all the particles that deposit in the body stay there, however, depending on how far into the respiratory tract they go; a respirable cyclone sampler is a suitable device for making these measures. In fact, inhalable sampling and respirable samples are common in occupational environments, where PM_{10} and $PM_{2.5}$ are measured for environmental exposures. Adding particle deposition in the airways and in the tracheobronchial and alveolar regions provides the total deposition as a function of particle size.

This analysis shows that almost all large particles and a large proportion of very small particles are deposited, but there is a clear minimum in the deposition of particles around 0.2 microns. Fortunately, said Koehler, this minimum overlaps with the common size distribution of a combustion aerosol, which means that a large proportion of those particles would be exhaled and not contribute to the dose.

To better understand what proportion of particles do stay in the respiratory tract, she and her colleagues developed a sampler that uses a piece of foam to serve as an inefficient filter—it captures only a fraction of particles that deposit in the respiratory tract (Koehler et al., 2009). It turns out that polyurethane foam has a similar open-cell, irregular structure as alveoli and, as Koehler explained, the mechanisms by which particles move and are lost through the different twists and turns of both systems are similar.

Her team tested this sampler in a cohort of stainless steel welders to assess exposures to nickel and chromium; exposure to these two metals is common among stainless steel welders, and both have important health consequences (Newton et al., 2021). Given their relatively short half-lives in the body, Koehler expected to see changes in dose over short-term changes in exposure. Urinary concentrations of chromium and nickel showed no association between the inhalable fraction of chromium and nickel and the urinary estimates of dose. However, examination of the lung-deposited fraction revealed a statistically significant association between the lung-deposited chromium or nickel and urinary biomarkers.

In closing, Koehler reiterated the complexity of extrapolating from particulate matter measurements to exposure, and the fact that exposure has different meanings according to where it is monitored and over what time frame. "For particulate matter, we might want to think carefully about the metric of exposure that we use…not just PM_{10} and $PM_{2.5}$, but thinking more about lung deposit or ultrafine particles that deposit at higher fractions."

THE UTILITY, USE, AND MISUSE OF LOW-COST CONSUMER INDOOR PARTICULATE MATTER SENSORS

There are a few important caveats in the field of low-cost, indoor particulate matter sensing, said Dusan Licina. The first is that this is a recent and rapidly growing field, with a majority of the published papers dating from the past 5 years. What this means, he said, is that what is relevant today might be outdated in just a few years. The second caveat is that more research is available on low-cost sensors used for outdoor particulate matter monitoring, and the result is that the type of monitoring networks established for outdoor air are not available for indoor particulate monitoring with low-cost sensors.

Licina then presented some definitions of terms he'd be using in his presentation:

- A reference instrument is associated with a federal reference or equivalent method, though many studies use laboratory-based instruments that are calibrated to function as a reference instrument.
- A monitor is an integrated device that comprises at least one sensor and other supporting components that create a fully functional air quality data collection system.
- A sensor is a subcomponent of a monitor that detects particles.
- A high-cost, laboratory-grade instrument typically costs between $3000 and $50,000.
- A low-cost monitor typically costs between $100 and $500, with a median price of around $200, and they mostly function as bulk particle detectors.
- A low-cost particulate matter sensor typically costs between $1 and $100.

Licina explained that there are different approaches to evaluating particulate matter sensors and monitors, and he noted that only about 10 percent of the published studies referred to a published protocol, such as the EPA's air quality specifications (Morawska et al., 2018). "Researchers typically adopt their own protocol for assessment of sensors and monitors, and this unfortunately results in variable judgment criteria for 'what do we consider to be good enough,'" he said. There is also great variability in methodology among research teams, including with regard to duration of testing, the environments in which they made their measures, the number of different technologies used, and the reference method they used. In addition, there is a range of factors to consider in terms of comparison with reference measurements, repeatability and reproducibility, limit of detection, dependence on particle size and composition, and dependence on indoor climate.

Assessing the Performance of Low-Cost Sensors

Researchers have found a generally high correlation between low-cost sensor readings and reference measurements, though such sensors typically perform much better in a laboratory environment than in the field (Demanega et al., 2020). "In the field, these sensors can suffer significant response changes, which can be attributed to changing conditions for particle composition, particle size, but also dynamic variable factors such as indoor climate," explained Licina. When comparing the use of these sensors in the laboratory versus a field setting, it is difficult to maintain a low particulate matter concentration over a long time in the laboratory, and the composition and concentration of a test aerosol in the lab may not be representative of aerosols in the study area. On the other hand, in the field it is necessary to account for variable particle composition, size, and changing environmental factors.

In terms of repeatability and reproducibility, most sensors demonstrate high intramodel consistency (Mukherjee et al., 2017; Zou et al., 2020), though these tests are usually performed in the lab, not in the field. Licina noted that repeatability and reproducibility could be influenced by the particulate matter concentration, range, source type, and drift. For example, the reproducibility for cigarette smoke is higher compared to Arizona test dust,[1] and most sensors detect higher concentrations of organic than inorganic particulate matter at identical concentrations because of differences in the absorption and scattering properties of these two types of particulate matter.

Low-cost sensors can have use issues related to their limit of detection, and their performance at concentrations lower than 20 µg/m^3 is often compromised (Holstius et al., 2014; Jayaratne et al., 2020; Jovašević-Stojanović et al., 2015; Kumar et al., 2015). Licina said it is important to calibrate sensors individually for each environment of intended use. His group in Switzerland, for example, has been field testing sensors in relatively pristine environments, where many of the sensors reported a value of zero for $PM_{2.5}$ concentrations. At higher $PM_{2.5}$ mass levels, low-cost sensors consistently report values that compare well to the reference value.

Environmental factors such as humidity and temperature also play a role in sensor output. Low-cost sensors, Licina explained, do not dry particles, unlike more traditional instruments, and this can compromise their

---

[1] Arizona test dust is a standardized dust that is used in filter and environmental contaminant tests.

accuracy due to particle hygroscopicity. Studies have found that humidity matters more than air temperature for sensor performance, which may be compromised at 85 to 90 percent relative humidity (Wang et al., 2015). In addition, a particle's composition can affect its hygroscopicity, so certain mixtures of aerosols could be more susceptible to relative humidity.

Suggestions for Stakeholders

Licina then provided a few suggestions for different stakeholders. For standards and guideline developers, he urged them to formulate guidelines for assessing short- and long-term performance sensors that everyone can use. He proposed that researchers standardize performance testing and allow comparisons between studies, and pretest and calibrate their sensors under the conditions in which the instruments will be used. For sellers and manufacturers, Licina's view is that they should offer a selection of sensors and monitors that are precalibrated for various types of indoor and outdoor environments, and provide more transparency regarding their calibration algorithms. For nonexpert users, he suggested making friends with an air quality expert who has experience with these sensors and monitors.

He then touched on several deployment challenges and needs. Currently, no local building, health, or safety code requires continuous monitoring of indoor particulate matter. The US Green Building Council's WELL Building Standard, a performance-based system for measuring, certifying, and monitoring features of the built environment that impact human health and wellbeing, does not require indoor monitoring of particulate matter, but it does offer an option to earn additional points on their rating system if one PM_{10} or $PM_{2.5}$ sensor is in place for every 325 square meters of space, a figure that Licina characterized as arbitrary. The following research questions may provide the data to develop more accurate and more precise guidelines:

- What are the best ways to ensure long-term sensor performance in field environments?
- What is the optimal time resolution for low-cost particulate matter sensors?
- What is the optimal sensor placement and density to capture human exposures?

Challenges also arise from the fact that indoor aerosols are episodic in nature and that indoor spaces are frequently subject to strong spatial gradients that are typically associated with humans and their activities. As a result, every person is enveloped in their own unique cloud of $PM_{2.5}$, and stationary indoor and outdoor monitors cannot capture that type of effect.

Solving that challenge will require developing portable, inexpensive, and robust continuous particulate matter sensors. This is a growing field, said Licina, and it will be important for improving the nexus between indoor spaces and exposures to particulate matter.

Other needs include developing low-cost sensors for ultrafine particulate matter, a challenge given that particles smaller than 0.3 microns do not scatter enough light and cannot be detected reliably using inexpensive optical methods. Licina noted that, depending on the source of particulate matter, low-cost $PM_{2.5}$ sensors may in fact detect agglomerated ultrafine particles. He also raised the questions of whether low-cost sensors need to be as good as high-grade equipment and whether sensor data can serve as a new class of measurements rather than a proxy for traditional measurements. He believes that the answer to the first question is "probably not" and that they need to go through further refinement. The answer to the second question is "yes," given that low-cost particulate matter sensors enable new types of observations that were not possible with traditional snapshot measurements.

Concluding his presentation, Licina noted that no currently available sensor is ideal for all applications, so it is necessary to find the optimal trade-off for the desired application. For indoor particulate matter management, low-cost $PM_{2.5}$ sensors are probably good enough, except in instances that require precise and absolute quantification. Regarding research needs, he singled out the need for more long-term field validation studies, as well as studies to assess the suitability of new sensors and monitors for health-related studies. "Advancing knowledge in low-cost measurement techniques for indoor particulate matter increases the likelihood that future control interventions can be used both to prevent undesired health consequences and to promote beneficial health outcomes," said Licina.

DISCUSSION

Moderator Elizabeth Matsui summarized her impression from the perspective of someone interested in the health effects of indoor particulate matter. This session, she said, highlighted opportunities to fill in information gaps identified in the day's first session by, for example, using low-cost air quality monitors to generate or refine indoor exposure data.

Howard Kipen observed that Nazaroff's presentation suggested that the chemistry of individual particles was of secondary concern when evaluating exposure impacts. Nazaroff responded that this is a judgment call and that reasonable people could disagree on that point. From his reading of the literature, his sense is that the physiologic response to particles is driven by the body sensing the particle as a foreign body and prompting an inflammatory and immune response. He wondered whether the chemical content of the small amount of material inhaled is sufficient to trigger what toxicologists would see as a response to a poison. Matsui explained that allergens or microbial agents such as endotoxins, which can be components of $PM_{2.5}$, can trigger adverse health effects through various immunologic and inflammatory pathways.

A workshop participant asked if there is value in using lung-deposited surface area as a metric for measuring ultrafine particles and comparing that with the foam sampler Koehler developed. Koehler thought that would be a useful comparison, but had not seen evidence that lung-deposited surface area is the "right" metric to use when trying to relate ultrafine particle exposures to health consequences. However, the devices used to make that measurement could prove useful for providing a better understanding of the spatial and temporal patterns in ultrafine particle deposition.

Rengie Chan asked the panelists to talk about the advances needed for low-cost particulate matter sensors to be ready for routine use in classrooms. Licina replied that some people in the field believe that the current crop of low-cost $PM_{2.5}$ sensors are ready, given that when it comes to managing indoor levels it is not necessary to know the exact concentration. It is useful, however, to know if the air quality is better now than an hour or day ago. Manufacturers of these devices are not always striving for higher accuracy but rather the best trade-off between a number of different goals, he noted. Nazaroff agreed and echoed Licina's idea of low-cost sensing providing a different category of data that would be useful for public messaging and public information. He would like to see research to identify the type of information that low-cost sensor networks can provide to help manage particulate matter levels in schools. "I do not think we are at the point yet where we can say we understand what these monitors can do such that we are ready to recommend to school districts that they buy a bunch of monitors and use them to help manage their indoor environmental air quality," said Nazaroff.

The speakers were asked what they thought about the use of a benchmark strategy for an indoor environment as a guide for identifying what constitutes good indoor air quality. Nazaroff replied that the indoor landscape includes public and private spaces where individuals have varying degrees of control over their environment and make decisions based on their own interests and knowledge; in contrast, everyone shares the outdoor environment and researchers and regulators can work effectively and collectively toward a common good. The indoor space has a mixture of individual actors making decisions based on their own interests and also their own knowledge, perceptions, or beliefs. This plays into the question of how to rationally devise an overall program to address indoor environmental quality, something that has been up for discussion for decades with no collective action at the governmental level.

One possibility, he said, is to look at other types of intervention programs in the built environment that help ensure safety, to provide guidance for moving forward on steps such as requiring regular inspections to ensure that ventilation systems are working properly and that filters are appropriately replaced. He also suggested that certification programs, and posting that information in a prominent manner based on an inspection protocol, could raise public awareness and trigger public cooperation. Koehler agreed and added that until low-cost sensors are of sufficient accuracy to be a compliance tool, perhaps the best approach is to provide concrete guidelines and recommendations to individuals and building managers on ways to improve air quality.

Asked about the long-term performance of low-cost, consumer-grade sensors and monitors, Licina said that this needs to be assessed, although manufacturers are improving their products on such a short timeframe that such tests are often outdated by the time they are completed. Regarding a question about how to convert particle counts to mass, some low-cost sensors do not measure levels but instead use algorithms to generate mass estimates. This is an issue that the research community is considering, to help ensure that these devices can be calibrated and adjusted to measure particles of different size and composition.

In response to a question about particulate matter exposure on weekdays versus weekends, Koehler acknowledged that there probably are differences. She also addressed an inquiry about whether the high concentration of small particles in rural areas could be soluble organic aerosols from trees or forests instead of resuspended particulate matter from roads, agreeing that this could be the case but adding that "these measures are mass based and there is very little ability to distinguish between the different sources."

A workshop observer asked the panelists to list approaches for identifying sources of particulate matter that have the most potent health effects, as a step to targeting those sources for control. Koehler replied that this is a difficult problem given the variety of potential sources and the data that Nazaroff showed regarding the short timescales over which particle concentration can significantly change. Nazaroff observed, though, that real-time monitoring, coupled with some combination of occupant diaries and real-time metadata measurement, does provide an opportunity to disentangle the contributors of indoor sources to indoor exposures without having to rely on chemical speciation of sources; this has been done to a limited extent and could be expanded to larger populations. The challenge, said Matsui, would be conducting those studies on a large enough scale, and Licina added that current technology struggles to characterize temporal, size-distribution, and compositional variability.

A final question from the audience asked the panelists to talk about the importance of measuring the ultrafine particle component of $PM_{2.5}$. Koehler replied that ultrafine particles represent an interesting exposure, but the sensors that differentially detect that fraction cost around $10,000 USD. For epidemiologic studies, this cost is prohibitive. Kipen added that ultrafine particles have proven to be a big challenge in the outdoor monitoring space because the numbers fluctuate very quickly, making it hard to use them to characterize an exposure. Their numbers vary more slowly indoors, so this may be less of a challenge in the indoor environment. Nazaroff pointed out that the building envelope does a good job of stopping ultrafine particles from infiltrating to the indoor environment and that indoor sources seem to dominate, which should make their assessment more feasible. Given the ways in which these particles can affect humans, he concluded that "ultrafine particles are another big world that is on the horizon in terms of public health concern."

7

Day Two Summary

To conclude the second day of the workshop, planning committee member Seema Bhangar summarized the key points of the presentations and discussions. First, she observed, the speakers had indicated that health burden from indoor particulate matter exposure is significant and is therefore a problem worth caring about and studying. "The speakers fleshed out in exquisite detail all the pieces of this, furnishing numbers of health burden, outcomes of concern, effect sizes, and mechanisms of effect," she said.

Second, measurement is important because it influences how health burdens are understood, attributed, and quantified. It also informs and validates control efforts. As Bhangar noted, knowledge about the health effects of particulate matter exposure is only as good as the metrics and measurements used to create that knowledge. In that sense, use of the right metrics and understanding of health effects have to continuously keep pace and bolster each other. In her view, the speakers made the case about the dynamic back-and-forth between conversations concerning health effects and metrics.

Third, measurement of particulate matter and health burdens is complex, particularly when it comes to addressing the question of what constitutes a "right," robust measurement method. Bhangar cited the need to articulate whether a particular health burden is acute or chronic, permanent or reversible, indirect or direct, and reported or measured using what metric or biomarker. In addition, there are the interactive effects that need study.

Fourth, the speakers identified a number of challenges to measuring exposure and dose and the possibility of mischaracterizing both. They also noted the challenges of developing and selecting metrics and measurement methods for different applications and environments.

Fifth, opportunities exist to use new measurement tools and methods to address those challenges and answer critical questions related to exposure and health outcomes. Almost every speaker, Bhangar remarked in closing, showcased those opportunities.

8

Indoor Particulate Matter Exposure Control and Mitigation

The final day of the workshop began with planning committee chair Richard Corsi posing three questions that would guide the day's presentations and discussions:

1. How do we reduce exposure to and health effects associated with fine particulate matter ($PM_{2.5}$) in buildings?
2. How do the decisions of occupants in their homes, or teachers in their classrooms, affect exposure to $PM_{2.5}$?
3. To what extent can filters and mechanical systems or standalone air cleaners reduce exposure to $PM_{2.5}$ and its health effects indoors?

The day's first session featured three speakers. Jeffrey Siegel (University of Toronto) discussed control of $PM_{2.5}$ levels in homes, and Elliott Gall (Portland State University) did the same for schools. Brett Singer (Lawrence Berkeley National Laboratory) then addressed how $PM_{2.5}$ exposure from cooking could be reduced. Following the three presentations, Rengie Chan and Seema Bhangar moderated an open discussion with the panelists.

$PM_{2.5}$ FILTRATION AND AIR CLEANING IN RESIDENTIAL ENVIRONMENTS

Indoor air filtration, said Jeffrey Siegel, involves more than just a device or filter. It comprises the totality of the filtration system and the context in which that system exists, for it is that context that determines everything about the performance of the unit. Context, he explained, includes the virus, particle, droplet, or contaminant that has to reach the filter, a trajectory that is influenced by air flow; the ability of the filter to remove the virus, particle, droplet, or contaminant, which is determined by the system's in situ efficiency; and removal that has to be big enough to compete with other removal processes in the environment. All of these determine the system's effectiveness.

Most homes in North America, said Siegel, have recirculating heating, ventilation, and air conditioning (HVAC) systems, which means they do not add outside air but only recirculate air already inside the building. Measured recirculation rates—the number of house volumes that go through the system when it is operating—span a wide range (Figure 8-1), which means that there is going to be a big difference in the performance of any filter depending on air flow (Touchie and Siegel, 2018). Another important parameter regarding air flow is runtime, the fraction of time that the HVAC system operates, and that varies, too, depending on the weather and other

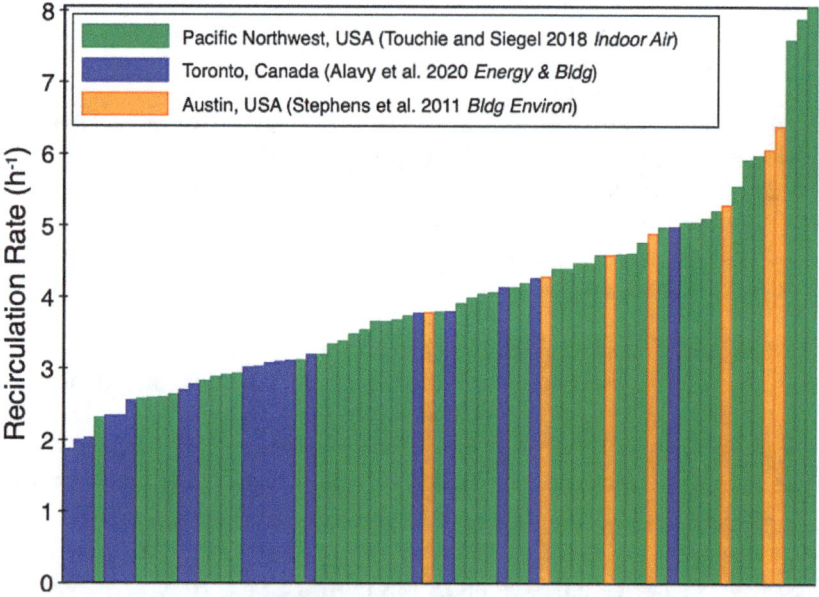

FIGURE 8-1 Home air volumes that pass through a filter when the heating, ventilation, and air conditioning system is operating in residences sampled in 3 studies in selected North American locations.
SOURCE: Siegel slide 4.

factors[1] (Figure 8-2). The median runtime is approximately 18 percent, which means that even with the best filter in the world, air will not be passing through it over 80 percent of the time.

Efficiency, measured by the percentage of particles that a filter removes in a single pass, can approach 100 percent for very small and very large particles (Hanley et al., 1994), which are removed by Brownian diffusion and inertial mechanisms, respectively, but it drops significantly for particles in the 0.1 to 0.3 micron range (Figure 8-3). Face velocity, the measured air speed at an inlet or outlet of an HVAC system, also affects efficiency, which drops as face velocity increases. "Right away, we see the beginnings of how operation might affect filtration performance," said Siegel.

There are several rating standards for filters. The standard set by the American Society of Heating, Refrigerating and Air-Conditioning Engineers (ASHRAE, Standard 52.2) is assessed as follows. In the laboratory,

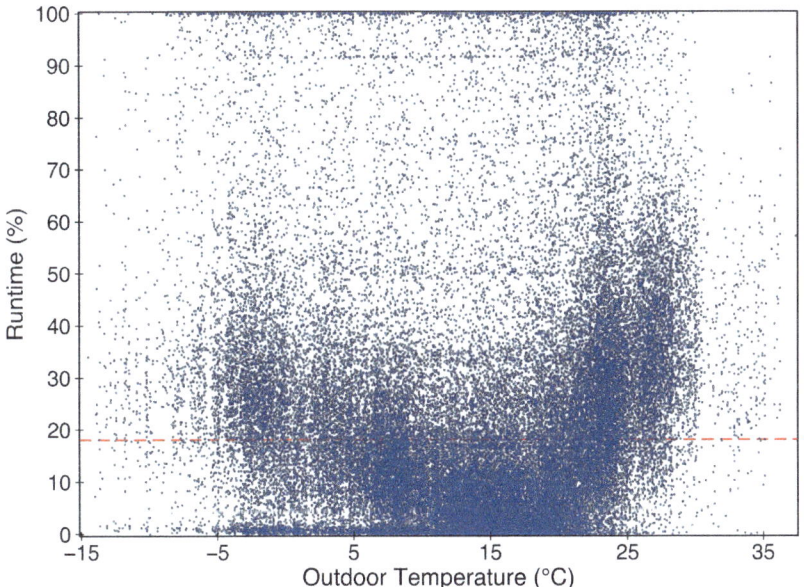

FIGURE 8-2 Fraction of time that a heating, ventilation, and air conditioning system operates.
SOURCE: Siegel slide 4, adapted from Touchie and Siegel (2018) Figure S6; reprinted with permission from John Wiley and Sons, Inc.

---

[1] Touchie and Siegel (2018) note that "[r]untime is influenced by climate and season but also by building characteristics, equipment sizing, occupant preferences, and operations and maintenance" (p. 906).

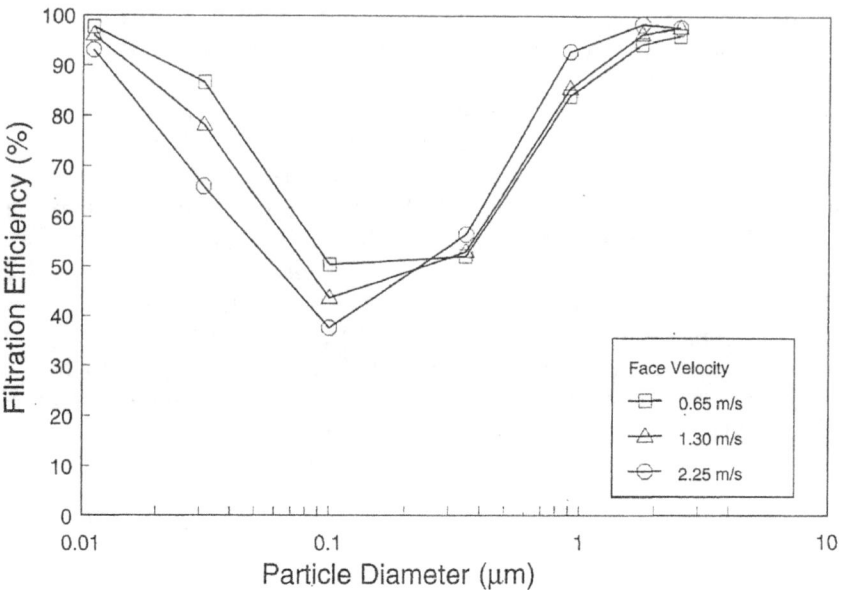

FIGURE 8-3 Single-pass filtration removal of particles at different face velocities.
SOURCE: Siegel slide 5, from Hanley et al. (1994) Figure 3; reprinted with permission from John Wiley and Sons, Inc.

a potassium chloride aerosol is run through a filter at fixed flow rate along with successive loadings of an artificial test dust. The test measures efficiency for three size ranges of micron—0.3 to 1.0, 1.0 to 3.0, and 3.0 to 10.0—and the results provide the minimum efficiency reporting value (MERV) for that filter.

Siegel noted three important points regarding this test. First, it is not relevant to many filters installed in residential buildings, but rather intended for filters used in commercial buildings; second, it is a laboratory standard, not an in situ standard; and third, it addresses only particles in the defined size ranges. In situ testing shows how various electret[2] and nonelectret filters perform in homes as opposed to in a laboratory setting (Figure 8-4).

The in situ testing results indicate that the MERV rating does not necessarily predict how well a filter will perform in a home, said Siegel. One reason for that is a phenomenon known as bypass, which can take several forms; it could be a large gap around a filter installed in an HVAC system, or a seating gap (space between the filter and the rack that holds it in place). In residential settings, it is common for HVAC systems to have a filter slot

---

[2] Electret filters are those whose filter media has been modified by the manufacturer to initially have an electrostatic charge on the fibers.

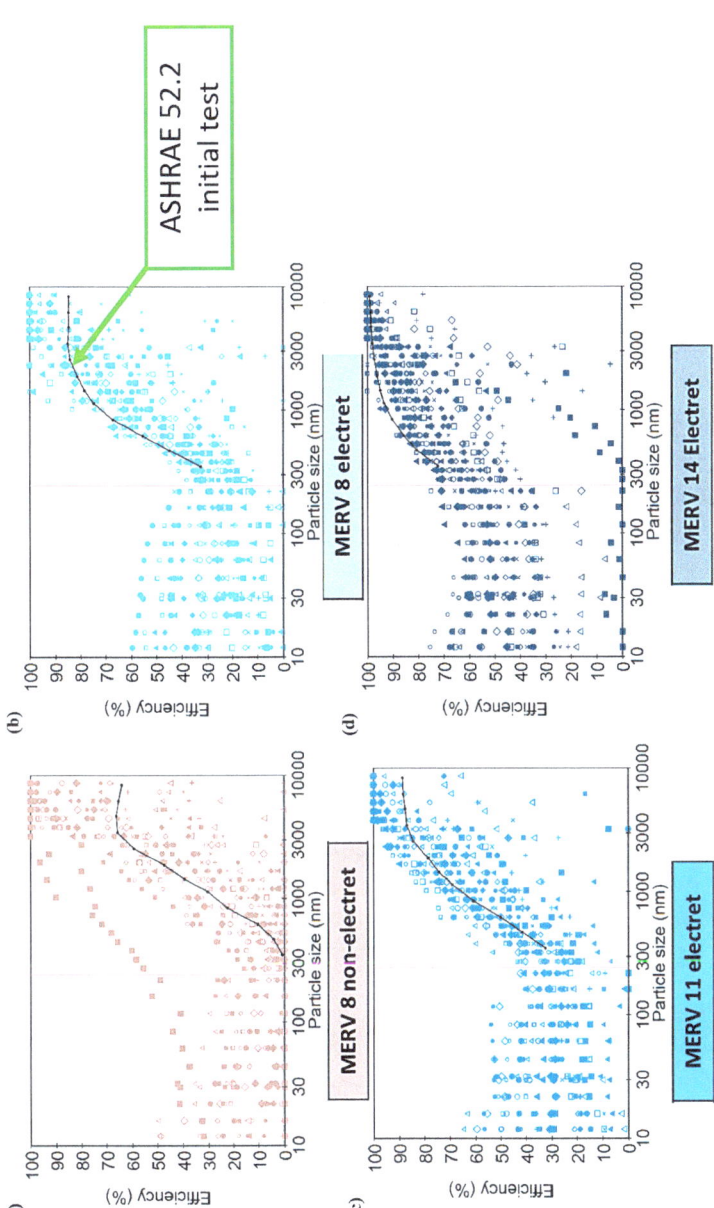

FIGURE 8-4 Results of in situ testing of four types of air filters compared to laboratory results (black line). MERV = minimum efficiency reporting value.
SOURCE: Siegel slide 7, adapted from Li and Siegel (2020), Figure 2; reprinted with permission from John Wiley and Sons, Inc.

that allows for easy filter changing, but that space also allows air to bypass the filter. Bypass always has a negative effect on efficiency, but how much of an effect depends on the filter and the bypass gap in the system (Chojnowski et al., 2009; VerShaw et al., 2009; Ward and Siegel, 2005).

Another issue is that filter performance changes with age, particularly for electric or charged media filters that account for about half of the US market (Lehtimäki and Heinonen, 1994). One set of experiments in a commercial building, for example, found that even after one week, efficiency began dropping for smaller particles, and after 36 weeks the filter's performance had dropped significantly (Lehtimäki et al., 2002). Siegel has found similar results for charged or electrified filters in homes (Li and Siegel, 2020).

He noted that data from numerous studies show that filtration performance varies substantially and follows no particular pattern (Alavy and Siegel, 2019). He explained that there are many reasons for this, including runtimes, bypass, and the characteristics of the building. For example, in a leaky building where there is already removal of particles by ventilation, filtration will not have much of an effect (Zhang et al., 2020). "To have effective filtration you have to have a good filter, it has to be properly installed, it has to have a big enough flow rate, you have to have a long enough runtime, the filter has to be changed frequently, and of course the big picture is you have to effectively compete with what else is going on in the building," said Siegel.

Noting that other speakers in the session would be talking more about portable air cleaners, Siegel did comment that the same issues apply to these cleaners. In addition, placement of a portable cleaner relative to particulate matter sources makes a big difference in effectiveness and performance (Novoselac and Siegel, 2009).

There are other air cleaning technologies available for consumers, including those marketed with terms such as photocatalytic oxidation or ionization. Based on the literature, Siegel said, there is no independent evidence of efficacy or effectiveness in real environments for many of these technologies, and in fact the parameters that have been evaluated point to poor performance by most of these devices. For some, such as those that use ozone or ion generation, there is the potential for harm related to byproduct formation.

Research Needs

In Siegel's view, one of the biggest research needs, given contextual issues, is to develop a way of assessing in situ filtration performance in homes, potentially with low-cost monitors of the sort discussed during the second day of the workshop. The goal should be to have methods that can ensure that filters perform as they should over time in real-life settings. He also would like to see how personal monitoring, rather than a central measure of performance, could assess how well a filter reduces exposure.

Research is needed as well on alternative control approaches that would maximize performance in the context of energy use and that would preserve filter longevity to reduce the cost of filtration. One idea would be to use a low-cost particle monitor as a thermostat-like control that would turn on the filter when particle levels reach a set value. Siegel also raised the idea of using artificial intelligence to enable predictive air cleaning. In fact, he and his collaborators have been working on just such an idea, and in early tests their model was able to predict high particle concentration events 2 hours ahead of time a little more than half the time, and changing the model to minimize false negatives enables it to predict high particle concentrations about 90 percent of the time.

Third on Siegel's list of research needs would be to understand what exposures filters have avoided in the environments in which they have been deployed. In some 60 to 70 articles, investigators have used what he called filter forensics to quantitatively link contaminants extracted from HVAC dust filters to time-averaged integrated air concentrations (Haaland and Siegel, 2017). This work has produced detailed fingerprints of the concentrations of particle-bound contaminants, including the SARS-CoV-2 virus, that collect on filters. This type of research could provide a way of exploring indoor concentrations of various pollutants as well as the effectiveness of filtration and other control strategies.

$PM_{2.5}$ EXPOSURE CONTROL IN SCHOOLS

Schools are a critical environment for a susceptible population—children—and 15 percent of schools, with 6.4 million students, are less than 250 meters away from a major roadway, an important source of particulate matter (Figure 8-5) (Kingsley et al., 2014), said Elliott Gall. Schools with higher percentages of Hispanic, Black, and Asian students have disparate exposures to roadway-generated particulate matter (Grineski and Collins, 2018), making this an important environmental justice issue (Gaffron and Niemeier, 2015) as exposure to elevated levels of traffic-related air pollution can adversely affect student health and cognition.

For example, research has shown that traffic-related air pollution exposure is associated with increased asthma diagnosis (HEI Panel on the Health Effects of Traffic-Related Air Pollution 2010) and with lower scores on measures of working memory and other cognitive markers (Sunyer et al., 2015). Gall said that some of the challenges in assessing and mitigating school-based exposures to elevated air pollution levels are that those levels

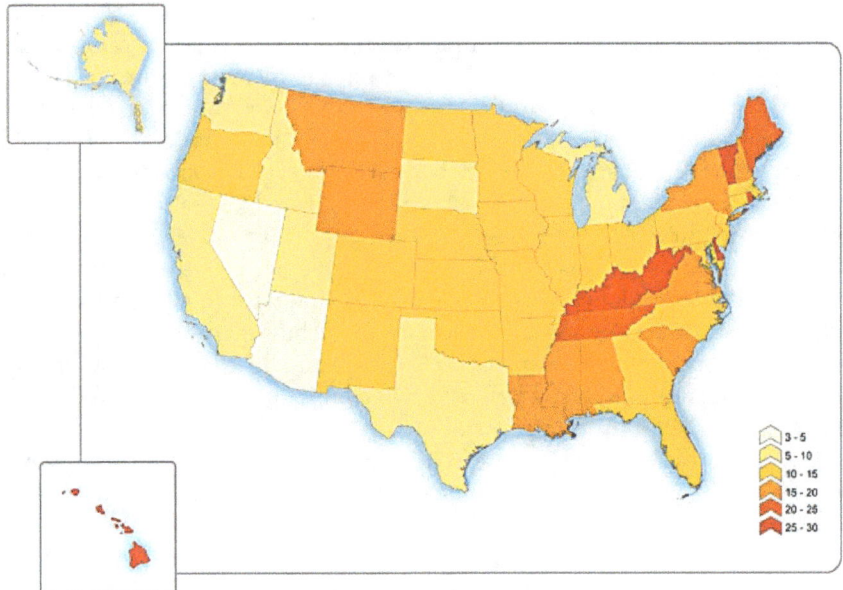

FIGURE 8-5 Percentage of students attending a school within 250 meters of a major roadway (2005–06 school year).
SOURCE: Gall slide 2, from Figure 1b in the cited publication.[1]

[1] Reprinted with permission from Springer Nature Customer Service Centre GmbH: Springer Nature. *Journal of Exposure Science and Environmental Epidemiology* vol. 24(3). Proximity of US schools to major roadways: a nationwide assessment. Kingsley SL, Eliot MN, Carlson L, Finn J, MacIntosh DL, Suh HH, Wellenius GA. pp. 253–59. © 2014.

may vary in space and that weather conditions can be quite important in determining those levels both around and in schools.

Over the past several years, Gall and the members of his research group have been studying traffic-related air pollution constituents in a middle school in Portland, Oregon, that was renovated in 2018 to serve Portland's historically Black community. The school is adjacent to Interstate 5, and measurements of black carbon, ultrafine particles, and $PM_{2.5}$ taken on the school's rooftop prior to the renovation showed elevated levels of these three types of particulate matter, with the exact burdens changing with wind speed and direction. For example, air movement across the highway toward the school led to large masses of black carbon and ultrafine particles. Gall clarified, though, that while there was some association between $PM_{2.5}$ levels and wind speed and direction, the extent to which PM levels changed as a function of these parameters was less than was observed for traffic-related air pollution constituents such as black carbon and ultrafine particulate matter.

This finding is important for two reasons, said Gall. Research suggests that, in general, the health effects associated with traffic-related air pollutants such as black carbon (Janssen et al., 2011) and ultrafine particles (Schraufnagel, 2020) are greater than those associated with $PM_{2.5}$. In addition, ventilation standards related to minimum acceptable indoor air quality, such as ASHRAE 62.1,[3] trigger increased filtration efficiency requirements only if the National Ambient Air Quality Standards are exceeded. Given that ASHRAE 62.1 references $PM_{2.5}$ levels and that the air in the school's region is in compliance for $PM_{2.5}$, there would be no increased filtration efficiency triggered by compliance with ASHRAE 62.1 unless an engineer or building designer identified a specific near-roadway source in a site survey.

Gall said that there has been a great deal of work on traffic-related air pollution in urban environments and that this work has shown that there are strong spatial gradients of traffic-related air pollutants in those environments. Within a zone of about 200 to 500 meters from a freeway, there are elevated levels of a variety of constituents associated with vehicle emissions (Karner et al., 2010). While that zone can expand to thousands of meters at night (Kozawa et al., 2009), that is not likely to be a major concern for schools. Some constituents of traffic-related air pollution, including ultrafine particles, black carbon, and volatile organic compounds, diminish rapidly within 150 meters of the highway, while others, including $PM_{2.5}$ and benzene, decay more gradually. "The opportunity here is that we can leverage this spatial gradient in traffic-related air pollutants in near-roadway schools to potentially reduce exposure," said Gall.

Mitigation Approaches in Schools

One straightforward mitigation approach is to consider placement of the school's fresh air intake as far as possible from the roadway. When monitors were placed on two faces of the school—one close to the freeway, the other 80 meters away and facing a residential street—the resulting data showed that relocating the school's air intake to be further from the highway would have an effect equivalent to installing a MERV-8 filter, but without the energy cost of a filtration system (Laguerre et al., 2020).

Another mitigation strategy would leverage both diurnal trends in traffic-related air pollution and spatial trends to reduce student outdoor exposures during lunch and recess. Gall and his students did a spatial mapping of ultrafine particle concentrations in a nearby park, where students

---

[3] Additional information is available at https://www.ashrae.org/technical-resources/bookstore/standards-62-1-62-2.

go during recess and lunch, and around the school site by walking around the area with a handheld condensation particle counter throughout the day. The data revealed that peak levels of traffic-related air pollution occurred during the morning rush hour and remained elevated until the midafternoon, suggesting that shifting outdoor activities until later in the day would reduce the students' exposures.

Gall reported that this approach has been explored in four schools in Ottawa, Ontario, using a "smart" ventilation system that brings cleaner air into the schools from 5:30 to 6:00 AM and then changes to recirculation until the students arrive. With this approach, the two schools that started at 9:00 AM realized a significant reduction in ultrafine particles and VOCs in the school.

A third approach involves active air cleaning for particulate matter in occupied schools. A small body of research has shown that filtration can be effective, but the results obtained in real life are variable. The interventions studied included filtration, improvements to the efficiency of filtration and the HVAC system, or deployment of standalone air cleaning and mechanical filters in classrooms. While some interventions had removal efficiencies of 80 percent or higher, the average removal effectiveness was between 40 and 50 percent (Gao et al., 2019; Jhun et al., 2017; McCarthy et al., 2013; Park et al., 2020; Polidori et al., 2013; Scheepers et al., 2015; van der Zee et al., 2017). Gall attributed the variability to factors such as the presence of indoor sources of particulate matter, how leaky the environment was, and the location of the intervention in the case of standalone air cleaners.

In the case of the Portland middle school, the contractor installed a single air handler to serve the entire school. This air handler was outfitted with an advanced air cleaning system comprising a MERV-8 prefilter, a MERV-16 filter, and functionalized carbon that targeted gas-phase species that might be elevated in a near-roadway environment. The goal of installing this system, based on the monitoring work Gall and his students had performed prior to the renovation, was to reduce traffic-related air pollution in outdoor ventilation air provided to the school to be at least equal to that of a school located in the urban background with a standard filter (Laguerre et al., 2020).

For this school, prerenovation measurements showed that the site had approximately five times the black carbon levels of the urban background when the winds were blowing across the highway and toward the school. Calculations showed that the contractor's design was likely to have a removal efficiency of at least 84 percent for black carbon, which would meet the targeted goal. In fact, a 1-week monitoring assessment found that the effectiveness of the installed system was approximately 85 percent and that it effectively decoupled indoor from outdoor air pollution levels.

As Gall noted, there are indoor sources of $PM_{2.5}$ matter that need to be addressed, especially when targeting $PM_{2.5}$ as opposed to the traffic-related air pollutants such as black carbon. Using a mass balance analysis, his group determined the contributions to particulate matter from the outdoor air fraction of supply air, the recirculation fraction of supply air from the building, and occupants and their activities. A large fraction of the $PM_{2.5}$ sources are occupants and their activities.

Chemistry in the School

Chemical reactions involving VOCs can be an important source of indoor particulate matter. In the Portland middle school, the carbon scrubber component of the air cleaning system appears to suppress levels of secondary organic aerosol precursors and thus formation of secondary organic aerosols. Monitoring showed that there is very little particle formation when the air handling system is on, but higher total particle counts in the indoor air than in outdoor air when the system is off. "In the middle of the night, when the school is unoccupied, we see higher total particle counts in indoor air than outdoor air, which we think is evidence of chemical reactions that are forming fine particulate matter," said Gall.

In conclusion, Gall said there are opportunities for reducing $PM_{2.5}$ exposures in schools near roadways. They include increasing the distance of the HVAC system's air intake from outdoor $PM_{2.5}$ sources, altering the timing of activities, installing air cleaning equipment, addressing indoor sources, and possibly managing indoor chemistry via air cleaning or source reduction.

Regarding research needs, he listed the utility of more data on the efficacy of installed interventions in school, lower energy and maintenance methods for ventilation and air cleaning, more information on the strength of particulate matter sources in schools, and studies on the health effects of exposures to particulate matter of indoor origin.

MITIGATION OF $PM_{2.5}$ EXPOSURES ASSOCIATED WITH COOKING

Brett Singer reminded participants that, as Marina Vance explained during the first day of the workshop, cooking generates particulate matter from the food being cooked and from some burners. When a gas burner is working well, it produces carbon dioxide, water, nitrogen oxides, some formaldehyde, and ultrafine particles. Electric burners also produce ultrafine particles, and cooking itself produces $PM_{2.5}$, a large amount of ultrafine particles, and irritant gases. Induction burners appear to emit many fewer ultrafine particles and no nitrogen oxides, in part because there is no very hot surface involved (Less, 2012).

The operational concept for mitigating particle exposure from cooking is ventilation, which may involve a venting range hood, vented over-the-range microwave oven, downdraft exhaust system, wall exhaust fan, and/or open windows. The idea, explained Singer, is to remove particles generated during cooking before they can mix into the rest of the air in the home. Vents that run air through a filter and then back into the kitchen produce only small reductions in particulate matter.

There are some standards and codes that require kitchen ventilation, of which ASHRAE 62.2 is the most prominent; others include those promulgated by the California Building Code, the Home Ventilating Institute, Energy Star, and the International Residential Code. The basic requirement of ASHRAE 62.2 is to have a range hood above the stove that moves air more than 100 cubic feet per minute (cfm) and has a sound level of less than or equal to 3 sones, which is a moderately annoying level, said Singer, noting that the standard was set several decades ago based on the equipment available at the time; ASHRAE is now considering whether the requirements should be updated to mandate quieter hoods. He described the International Residential Code, which was created to enable local jurisdictions to set local requirements, as "basically worthless" for kitchen ventilation, with the result that US homes are still being built without kitchen ventilation installed or with ducting to enable the homeowner to have it installed later. "That is probably the single biggest problem here, because all the other problems I am going to talk about with the quality of the range hood and users actually activating the range hood, none of that is possible unless you have the device installed in your house, and that is very expensive if you do not have ducting installed for it," said Singer.

The effectiveness of a range hood is measured by its capture efficiency, which is the fraction of pollutants released at the cooktop or oven that the hood removes before it mixes with other air in the home. Singer explained that capture efficiency is calculated using carbon dioxide released from gas burners (or alternatively, a tracer gas), but a different approach is needed for particulate matter. His group has tried to quantify capture efficiency in the lab and shown that it increases with airflow and is much better for back burners (Lunden et al., 2015; Singer et al., 2012). In fact, for front burners, a range hood operating at 100 cfm captures only about 30 percent of the emitted particles. Over-the-stove microwave range hoods can be installed as either venting or recirculating, and when installed they need to be set to vent. If they are, tests by Singer's group found that they work as well as comparable range hoods. In his opinion, they are a good option if they are used.

Initial work on range hood capture efficiency focused on gases, which begs the question of whether the capture efficiency for them is the same as for particles. Singer's group attempted to answer that question and found that test results varied by a factor of two even when using consistent food sources and preparation time. Nonetheless, the results showed that range hoods effectively removed particles when cooking on the back burner and were less efficient at capturing particles released when cooking on the front burners. "When capture efficiency is high for gases, it is high for particles," he said. Data from his group showed, not surprisingly, that more particle capture occurs as airflow rates increase (Sun et al., 2018).

In one set of experiments, Singer and his colleagues measured gas and particle levels in the kitchens and bedrooms of nine homes, six of which had range hoods of various types (Singer et al., 2017b). The installed range hoods provided varied levels of exposure reduction, with the best performance seen with a hood that extended over the entire stove. However, the presence of a good range hood does not guarantee that the occupants of the home will use it, and a Web-based survey or more than 2000 people living in Southern California homes built between 2003 and 2010 found that about three-quarters of those in homes with a kitchen range hood that exhausts outdoors used their hood sometimes, most of the time, or always, whereas hood use was less in homes where the hood vented back into the kitchen.

When asked why they did not use their range hoods, the majority of people said they did not think it was necessary. A surprisingly large number said their range hoods did not work. In a subsequent study, Singer and his colleagues looked at range hood use in California houses and low-income apartments and found that occupants were more likely to use the hood when cooking for longer periods, but that, overall, most homes used a hood less than half the time they used their stoves or ovens. Research by a Canadian team found similar results (Sun and Wallace, 2021).

One of the big problems Singer said he has observed with range hoods is that the actual airflows in practice are much less than their product certification test result. He blamed this on the higher static pressure and more airflow resistance in homes compared to test chambers.

In conclusion, Singer noted that there are resources and guidance for increasing range hood efficacy (Phillips, 2019). His simplest advice was to use low-resistance ducting,[4] use a range hood that can move 250 cfm, make sure the range hood is quiet at reasonable air flows, and cook on the back burner.

---

[4] Phillips (2019) notes that considerations also include whether range hood ductwork is installed properly and whether the diameter of the ductwork is sufficient to allow the unit to operate efficiently without generating unacceptable noise levels.

DISCUSSION

Moderator Rengie Chan invited questions from the planning committee and workshop participants. Singer led off by asking Siegel if the key to using an electret filter is to replace it more frequently, and if so, how frequently. Siegel replied that in principle, replacing filters more frequently is a great idea. "In fact, if I could get people to change their filters we would be a long way toward being in a better place," he said. The problem with electret filters, he continued, is that performance degradation can happen within a few hours depending on the loading, and many of the homes in his study had less than 100 hours of runtime. What he would rather see than having people replace their filters twice as frequently is to invest in a sealed filter rack that eliminates bypass.

Noting that bypass is common in both commercial and residential systems and a problem that is well known to both researchers and practitioners, Chan asked Siegel to comment on what is needed to move to better-quality installations. Siegel replied that he has addressed bypass in every home in which he has lived and has been surprised at how inexpensive a problem it is to address. Consumer and installer education is what is needed, he said. Singer commented on the cost of a decent MERV-13 filter, and added that getting consumers to actually change their filters is still an issue. Gall noted that the main concern of the Portland school districts he studied was the cost and recurring maintenance associated with the installed HVAC system.

Chan then asked Siegel if the guidance on using MERV-13 or higher filters as a mitigation measure for the SARS-CoV-2 virus is a useful strategy. He replied that since the virus is much smaller than many of the particles he studies, in theory filtration should be an effective mitigation strategy. However, if the virus is part of a larger droplet, then the answer depends on how efficient a MERV-13 filter is at capturing those larger aerosols. His understanding is that most virus transmission involves 1- to 3-micron particles and a MERV-13 filter would have an 85 percent single-pass efficiency for that size, which suggests that filters can be an effective part of a mitigation strategy. He added that published results show that filters capture RNA viruses and sometimes DNA viruses. The problem is that if the HVAC system is running only 10 percent of the time, then 90 percent of the time it does not matter what the MERV rating is.

An audience member asked Gall if there is a level of airborne black carbon that is high enough that schools should not allow outdoor activities. The short answer, he replied, is no, there is not a specific number at which you might confidently say, 'Do not let children outside to play in the park adjacent to the school.' He noted that there are no federal standards for ambient black carbon levels, and, while Oregon does have a benchmark, it is based on the risk of one additional cancer in a population of one million people and does not address the question of the incremental risk of being outside when black carbon levels are elevated. He did point out that there are data showing that short-term exposures to traffic-related air pollution, even on the order of minutes, have been associated with certain cardiovascular metrics such as increased heart rate. But, he added, "it becomes quickly a very political process to say where students can go, when can they go there, what is outdoor recess going to look like at a near-roadway school."

Asked how the Portland school fared during Oregon's bad wildfire season in 2020, Gall said that he did not have intensive monitoring then, but there were low-cost particle monitors in the school that indicated that the HVAC system did a good job of keeping $PM_{2.5}$ levels low inside the school during the wildfire event. In fact, there were discussions with the school district and local county health department about the possibility of using the school as a clean air shelter for those who might seek to reduce their exposure during a wildfire. The COVID-19 pandemic derailed that idea, but it might be reconsidered in the future.

Singer brought up the issue of equity and the challenge of air cleaning in places that do not have mechanical cooling, which is still the case in many schools around the country. Gall replied that this is a significant issue in Portland, which historically has a mild climate and so every school does not have a mechanical cooling system installed. In fact, he said, at least one published study reported that an intervention was less effective than designed because people were opening windows to provide ventilation and cooling during hot weather, which substantially reduced the effectiveness of the standalone air cleaning system being tested. At a Portland middle school that was the subject of an intervention, the windows facing the highway were modified to no longer be openable to reduce infiltration from the freeway. He noted, though, that this school was the focus of a years-long effort and a substantial investment to improve the air quality in that one school. "This is a major challenge that school districts are going to face," said Gall. Singer added that the Environmental Protection Agency's Tools for Schools program is a good source of information for schools that are concerned about their indoor air quality.

Kim Prather commented that she has been helping many schools reopen after the COVID-19 pandemic forced them to close and that the most common mitigation procedure has been to install plexiglass barriers to prevent viral transmission. Siegel responded that while plexiglass barriers protect against very large droplets, those droplets are not responsible for most infections, so plexiglass barriers do not address the big risk. He affirmed the importance of physical distancing and people management.

Stuart Batterman (University of Michigan School of Public Health) noted that schools often have HVAC systems with filters that are hard to get to and change, and he wondered if dedicated systems could provide better filtration for outside air. Gall said that the mechanical contractor for the Portland school he studied did look at such a system, since it allows for a dedicated outdoor air supply with air cleaning that would have addressed the issue that school was facing. He was not privy to the discussions of why the school decided on a single air handler, though it does offer the advantage of cleaning human aerosols involved in disease transmission. Siegel remarked that there is a need to develop systems that require less maintenance given the maintenance gap that already exists in many schools.

Elizabeth Matsui worried that disparities will be exacerbated without a solution to address the indoor air quality of schools that are in poor communities and disproportionately in communities of color. Singer agreed and thought that an important step in the right direction would be for school districts to use money in the American Rescue Plan Act of 2021 (Public Law 117-2) designed to support school facility repairs and upgrades. In fact, he said, schools recognize the importance of providing ventilation and this is one challenge that can be addressed at a relatively low cost by improving the filter compartment, using a good filter, and maintaining the system properly. "This is a relatively small lift to potentially greatly reduce exposure for some of the most vulnerable people in our society," said Singer. Gall added that it took a confluence of community interests to put potential solutions forward for the Portland middle school and force the school board to consider them and act. "There certainly needs to be that ecosystem of awareness and action to bring these issues to attention and to have them acted upon," said Gall.

Turning to the subject of range hoods, Chan asked Singer to comment on the idea of phasing out gas stoves as a solution to the indoor particulate matter issue. Singer replied that getting rid of gas stoves would eliminate one important source of ultrafine particles and nitrogen oxides, though it does not remove the need for good ventilation. He commented that research that he and others have done has shown that gas stoves, especially in smaller homes, can produce high exposures to nitrogen oxides. He also noted that when it comes to making quieter range hoods, manufacturers are going to continue to charge more for quieter models until they are required to produce them and stop making noisier models.

When asked to talk about the type of low-cost particle monitors they are using, Siegel replied that he uses several different models, and he encouraged the audience to review Dusan Licina's discussion of various issues with these monitors. "It is important to understand the monitor you are using and its response, and I do not think that low-cost monitors eliminate the need for using more sophisticated instrumentation, but I certainly think that they have a huge role in helping us improve the performance of filtration," said Siegel. Gall agreed and noted that the low-cost monitors in the Portland middle school during the wildfire event provided some assurance that the system was working and could convince people that the air was better in that environment. Singer added that a test method for low-cost particulate matter monitors is making its way through the ASTM[5] process and his hope is that when it is produced it will allow a consistent standard to be applied to these monitors. While this standard will be imperfect—as all standards are—it will serve as a general reference and help consumers distinguish between products that work fairly well from those that do not work at all.

---

[5] ASTM is an international testing and standards organization.

9

Occupant Responses to Indoor Particulate Matter

The workshop's final session focused on how human behavior and public health considerations inform indoor particulate matter exposure mitigation strategies. The session featured three speakers. Stuart Batterman (University of Michigan School of Public Health) discussed human behavior as it relates to the use of portable indoor air cleaners, and Lindsay Graham (University of California, Berkeley) spoke about how building occupants interpret and respond to data from indoor air quality monitors. Sarah Coefield (Missoula City-County Health Department) then explained the role of public health in reducing community exposure to fine particulate matter ($PM_{2.5}$). Following the presentations, Rengie Chan and Seema Bhangar moderated the final open discussion of the workshop.

PORTABLE INDOOR AIR CLEANERS AND HUMAN BEHAVIOR

Stuart Batterman, who has conducted a number of field studies of indoor environmental exposures and health, began by providing the typical guidance he gives people who have children in Detroit who have asthma:

1. Do not smoke indoors or allow others to smoke indoors.
2. If you have a portable air filter, put it in your child's room and use it all the time. People who turn it on and off tend to leave it off for long periods of time, which reduces its effectiveness.

3. Instead of rags, use a microfiber furniture duster when dusting. It removes dust better than ordinary rags, which tend to spread dust around rather than pick it up.
4. To clean floors, use a vacuum cleaner instead of a broom. If you need to use a broom, sweep gently. Vigorous sweeping can throw dust back into the air and under furniture and appliances.
5. If you have forced-air heat or central air conditioning, use a good furnace filter to reduce particulate matter. A standard furnace filter (costing $1–5, often colored blue) does not improve air quality. Look at the minimum efficiency reporting value (MERV) number, which is the efficiency of the filter for removing particles from the air; a filter with a MERV rating of 13 or greater is generally a good choice and costs $15–60. Change filters every 3 months.
6. Do not use air fresheners and minimize use of mothballs and deodorizers.

He also advises people to change certain behaviors to improve their environment, their health, and the health of their children. In the public health world, multilevel interventions or multilevel approaches to changing behavior are known to be most effective; individual-, community-, and national-level interventions work best when applied together. At the individual level, this would involve identifying individuals at risk, such as children with asthma. Community-based interventions aim to modify the environment and use peer influence, including the media, community-based screening, community-based organizations, and sometimes rules, restrictions, and taxes to motivate individual behaviors. Often, community health workers bridge the individual and the community to try to change behavior. At the national level, messaging, regulations, codes, and other actions are the levers to induce change.

These behavioral changes, said Batterman, apply to many types of decisions about things such as indoor and outdoor emissions and pollutant levels, exhaust fan operation, HVAC systems and ventilation rates, and the purchase and operation of filters. A systems approach should be taken for both buildings and health, bearing in mind that each indoor environment, household, and community is unique. "A major take-home here, though, is that we need to identify effective ways to change behavior," he said.

Drawing from the literature on interventions, Batterman developed a conceptual diagram for thinking about behavior change to improve health by reducing particulate matter levels using filters (Figure 9-1) (Michie et al., 2018). Achieving that goal, he explained, depends on behaviors that can be triggered by an intervention and recognizing that the effectiveness of the intervention depends on the context, the population, the extent to which intervention information reaches different parties, and the extent

FIGURE 9-1 Conceptual diagram showing elements of behavior change to reduce particulate matter (PM) exposure and improve health.
SOURCE: Batterman slide 5 (adapted from Michie et al., 2018, Figure 2).

to which individuals engage in changing their behavior to meet that goal. The diagram, while simplified (it omits feedback loops, multiple behaviors, and the time dimension), does represent the fact that changing behavior is a multilevel indirect and complex approach—and that it often fails. Nonetheless, there have been successes, such as getting people to stop smoking indoors, which took many years to accomplish.

Portable Air Cleaners Have Their Uses

The preferred option regarding air cleaners is to have one installed in the central HVAC system, but portable air cleaners do have their uses, such as when a space does not have a filtration option, as in a home with radiators or baseboard heating, or when upgrading to in-duct filtration is difficult, impossible, or too costly. A third reason, said Batterman, is when additional cleaning is desired to reduce risk, such as when an installed forced-air HVAC system is operating less than 25 percent of the time, when there is inadequate airflow to a particular space, when there are local sources, or for susceptible individuals. Portable air clearers might also be useful for removing gaseous and biological contaminants that would not be removed by filters in a central HVAC system.

Based on theory and experimentation, portable air cleaners can reduce particulate matter exposures substantially, and some studies suggest they can reduce the frequency of asthma symptoms. In practice, though, they often do not live up to expectations, and Batterman listed some of the reasons why:

- A filter's "clean air delivery rate" and particle removal capacity may be inadequate for the space involved.
- Air change rates and open windows can affect performance and are rarely measured.
- Performance evaluations have limitations, such as limited monitoring, no control over emission sources, uncontrolled seasonal variation, and small sample sizes.
- Placement, bypass, filter deterioration, and other factors that arise in real-life applications are not considered in performance evaluations.
- Filters are effective only if used and if they have high runtimes.

In a study that Batterman and his colleagues did in a low-income community in Detroit, they provided 89 families with HEPA filters fitted with a data logger to capture use history, cash to run the filters in the bedroom of a child in these households who had asthma, and community health worker home asthma education visits (Batterman et al., 2012, 2013; Du et al., 2011; Martenies et al., 2018). Another 37 families, serving as the control group, only received community health worker home asthma education visits. Batterman noted that many of the homes were not air conditioned in the summer, and 42 of the 89 families also received an air conditioner to control ventilation. Measurements of particulate matter in the houses with air filters showed that they significantly reduced particle number counts in the submicron and 1- to 5-micron size ranges, with reductions in the children's bedroom of 50 to 80 percent during a one-week assessment at the beginning of the study. The HEPA filters were rated at 330 cfm and had four fan speeds, and they were large enough that, even running at the lowest, least noisy speed, they provided a good number of air changes in the spaces in these homes.

For the entire study, which lasted through all four seasons, the results were mixed (Figure 9-2). Compared to the control group, the intervention groups had 50 to 91 percent lower levels of particulate matter in the children's bedrooms, which Batterman characterized as an effective reduction. When he and his collaborators monitored how the filters were actually used, they found that behavior was a substantial source of variability in the study. In particular, they found that the filters were used about 97 percent of the time at the start of the study, but that dropped to around 70 percent over the rest of the study, which he thought might be due to "a novelty

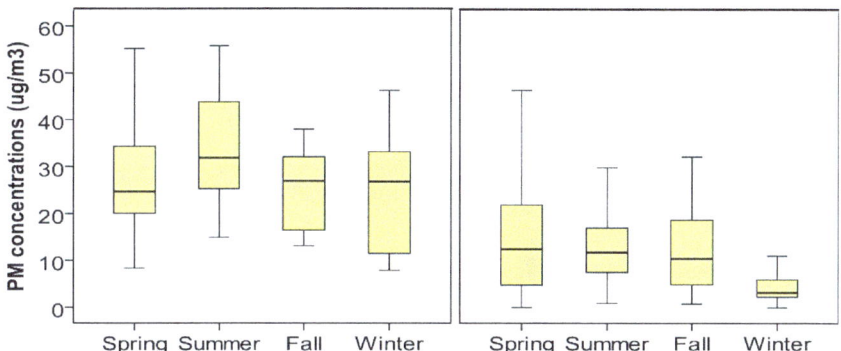

FIGURE 9-2 Seasonal summary of particulate matter (PM) concentrations for filter use study control (left) and intervention (right) groups.
SOURCE: Batterman slide 11.

effect." Moreover, some households used the filter 100 percent of the time, while others used it only 10 to 20 percent of the time.

One surprise was that filter use plummeted during the 3- to 5-month spans between monitoring periods, averaging only 28 percent of the time (Figure 9-3), and variability of use among the participants was dramatic. Batterman called this finding problematic and offered several possible explanations for the precipitous decline in use. One possibility is that the participants discounted the benefit of the filter, or that they only turned the filter on when they knew they were going to be visited and study staff expected them to be running the devices. For some households, the noise and draft associated with the filters were bothersome. There was an increase in electricity rates during the study, and though the study subsidized the cost of operating the filters, households may have cut back use when their electric bills rose. Batterman noted that operating a portable air cleaner can add over $200 to a household's annual electric bill.

Overall, the findings from this study showed that portable air filters can work when they are used, but effectiveness depends on their use, and in particular, on the user's behavior. In this case, Batterman characterized the filter use as "unexpectedly low."

The other part of the study involved examining the children's asthma symptoms and some biomarkers of inflammation. There was no significant effect on either, perhaps because of the low runtimes for the units. The lack of filter use, said Batterman, causes what is known as exposure misclassification, though he added that the indoor environment is just one of multiple exposure compartments.

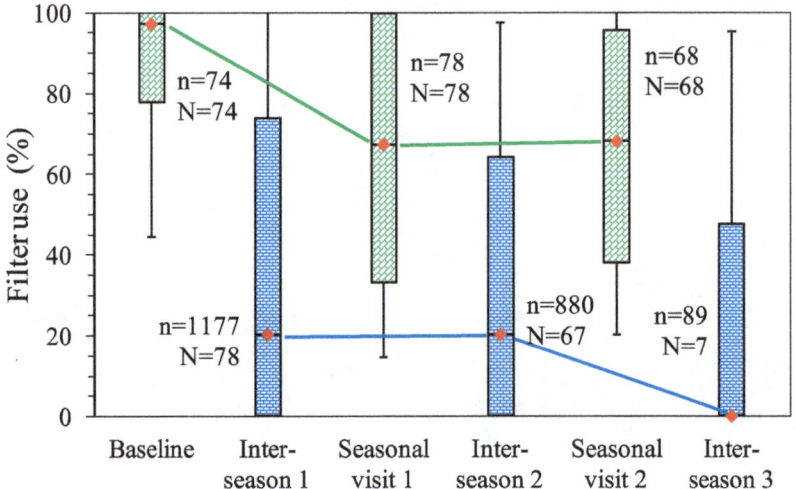

FIGURE 9-3 Occupant use of an air cleaner (filter) during and between monitoring periods. n: total number of weeks households were observed; N: number of households observed.
SOURCE: Batterman slide 5, adapted from Figure 1 in the cited publication.[1]

[1] Reprinted with permission from Springer Nature Customer Service Centre GmbH: Springer Nature. *Air Quality, Atmosphere & Health* 6(4). Use of freestanding filters in an asthma intervention study. Batterman S, Du L, Parker E, Robins T, Lewis T, Mukherjee B, Ramirez E, Rowe Z, Brakefield-Caldwell W. pp. 759–67. © 2013.

Modeling Filter Performance

As a follow-up to this study, Batterman and his collaborators used indoor air quality models to estimate indoor exposures to ambient $PM_{2.5}$, determine an "equivalent" exposure concentration that considers time and activity patterns, estimate the health benefits of using filters in schools and homes, and calculate the marginal costs of increasing filter use (Batterman et al., 2021). They applied the model to classrooms and homes with forced-air systems with filters installed in a child's bedroom and the living room. Overall, the model estimated that $PM_{2.5}$ exposure accounts for about 6.5 percent of asthma outcomes in children in the Detroit area. It also estimated that installing filters in classrooms would produce an 8 to 17 percent reduction in annual asthma burden, and in the homes of children with asthma would reduce the annual asthma burden by 11 to 16 percent. The marginal cost for a standalone filter unit was predicted to be nearly $500 a year, accounting for the initial cost, electricity, and filter replacements, and between $151 and $175 for a forced-air system in homes and $40 to $60 per classroom per year for schools.

In summary, Batterman said that providing portable filters should be considered an active intervention requiring behavioral change and multilevel interventions that address both the individual and community. His group's experience indicates that filter use or runtime should be monitored in trials to reduce exposure misclassification and that low runtime may help explain some of the variation in results that other studies have seen.

Research Needs

Turning to research needs, Batterman cited a need to understand what makes for a successful behavioral intervention, not just for portable air cleaners but for range hoods, exhaust systems, and other aspects of particulate matter control. While the literature indicates that information and simple messaging are important, they are no guarantee that people will change their behavior. The exception to this seems to be when the message permeates so deeply that it cannot be ignored, in which case behavior will change. "We are not at that point yet with the understanding about particulate matter indoors," said Batterman. But it is possible to facilitate behavior change by considering externalities, such as if an individual perceives that their behavior will help or hurt others or through peer pressure.

There are also questions about how to engineer portable air cleaners to encourage appropriate behaviors among users, make the cleaners more acceptable, and increase their use. This may require reducing their size, cost, noise, and the drafts they produce. Providing information and enhancing the public's awareness regarding filters might encourage people to buy them, use them, and engage in good behaviors. Some of this might be done, said Batterman, by incorporating both use or runtime information and indoor air quality sensors in air cleaner units and making that information intelligible for the user.

A final research need is to improve the controls for portable air cleaners. Running them continuously is one option, but using automatic, smart, or timer-oriented approaches or incorporating a particle monitor and occupant sensor would provide a number of advantages compared to manual controls.

HOW BUILDING OCCUPANTS INTERPRET AND RESPOND TO INDOOR AIR QUALITY SENSOR DATA

The occupants of a home or building have the power to affect indoor air quality in their environment, said Lindsay Graham, so trying to understand how individuals think, feel, and behave in relation to the information they receive is important for being able to shift behavior changes in a positive way. She showed information from a large database her research group has created from the Center for the Built Environment Occupant Survey,[1] a postoccupancy evaluation survey representing 90,000 respondents and nearly 1000 buildings worldwide (Figure 9-4). These data show that "occupants realize that air quality may be influencing them for better or for worse, but we are split on how those perceptions really fall out."

When she and her team looked at why people were dissatisfied with indoor air quality, the biggest reasons were tied to odors, with food tending to be the biggest source of dissatisfaction, though other factors such as odors from furniture and cleaning products were also important reasons. Research has shown that occupants often include temperature in their assessment of air quality (Melikov and Kaczmarczyk, 2012; Schiavon et al., 2017). In fact, people often conflate the idea of positive air quality or "freshness" with air temperature, air speed, and humidity, which indicates that there are many factors that determine how people feel and think about their air.

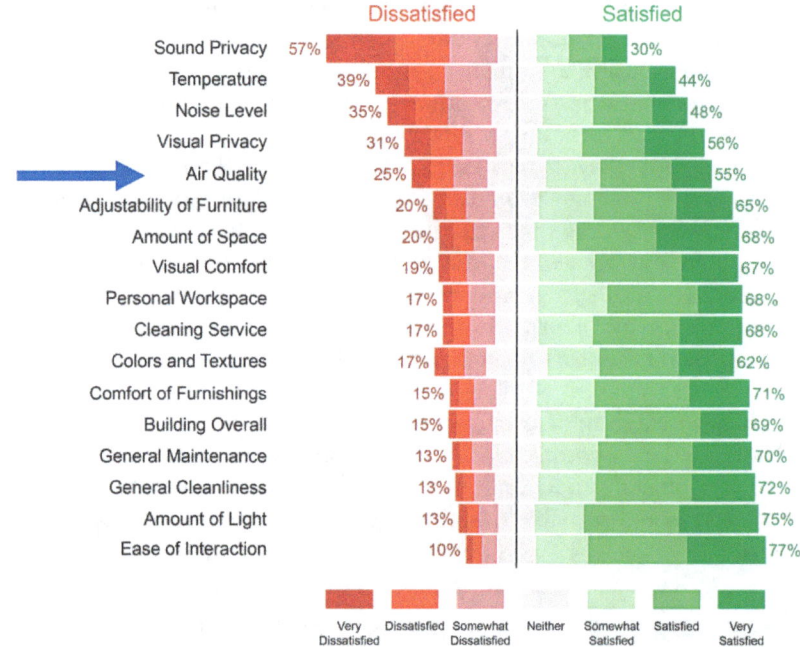

FIGURE 9-4 Occupant perception of indoor environmental quality, including air quality.
SOURCE: Graham slide 3, adapted from Graham et al. (2021) Figure 3; reprinted with permission from *Buildings and Cities*.

---

[1] https://cbe.berkeley.edu/resources/occupant-survey/

"It also tells us this is something that is hard to detect" by an individual, said Graham—air quality is not something an occupant can see, touch, or feel without the aid of a device that provides a metric. That creates a problem for getting people to pay attention to their behaviors that affect the air quality in their environments. When communicating about air quality, much of what is known comes from how outdoor air quality is reported, and over the past few years there has been an increase in efforts to raise awareness about outdoor air quality.

As Batterman noted, addressing how to effectively influence behaviors with information is an emerging area for indoor air quality. Graham and her collaborators have begun to explore that at an individual level, attempting to identify what type of information occupants prefer when trying to interpret indoor air quality and what information will lead to easy and accurate interpretation of what the indoor air quality of a space might be. To do this, they identified common environmental quality data visualization tools, generated and tested three different types of data visualizations—"numeric" (data reported in a table), "scale" (presented as points on graphs), and "health" (via icons representing population groups and states of health)—and evaluated them in an online survey with almost 250 participants. Each of the three visualizations represented the same scenario depicting high $PM_{2.5}$ and elevated CO_2 levels that might be typical in a home during cooking or cleaning.

The results of this experiment showed that participants preferred the "health" visual, though they judged all three to be easy to interpret. "What this told us was that perhaps it is not about the complexity of the information given and one's ability to interpret the information, but perhaps preference is driven by the content itself," said Graham. To understand this in a deeper way, she and her team examined the qualitative data they collected and found three broad patterns: participants preferred information that was easy to understand and interpret, that provided visual or graphic information, and that provided actionable information. What the participants did not like in a visualization was complexity—graphics that were confusing, not clear, not intuitive, and not easy to understand. The also did not like when action items were not included or when there was little to no information on the cause and effect of pollutants. Graham inferred that "occupants are hungry for information to try to understand and educate themselves around why air quality may or may not be helpful to them in that moment."

Regardless of the visualization type, participants who believed the indoor air quality to be poor were more likely to also believe that they needed to take action. However, the participants were more likely to believe the air quality was bad and the environment unhealthy when viewing the "health" visual. "This shows us that there is this intuitive linkage between understanding that the environment is unhealthy and the need to try to move toward a healthier space, indicating that there may be some motive there that we could utilize when trying to depict information like this to occupants to nudge their behaviors," said Graham.

One of the goals of this work, she explained, was to help create guidance on how to depict information for the general public regarding indoor air quality in a way that would motivate them to take action. She recommended providing occupants with clear, graphical, and actionable information relevant to their health and behavior, not just the air, and outlining steps to overcome poor indoor air quality in their spaces.

For research needs, Graham said it is important to keep working to understand individuals and what drives their behavior—to determine whether there are individual differences, such as personality type, gender values, motivations, and lifestyle that affect data interpretation. She would also like to see research on whether perceptions translate into occupant actions, and efforts to identify building or behavioral interventions that could be implemented to nudge people toward engaging in healthier behaviors. As a final comment, she emphasized that occupant education, engagement, and empowerment are paramount when seeking to foster actions to improve indoor air quality. "Without that educational piece, we do not give occupants the potential to help us craft healthy and effective spaces where we can thrive," said Graham.

PUBLIC HEALTH RESPONSES TO REDUCE COMMUNITY EXPOSURE TO INDOOR $PM_{2.5}$

In the workshop's final presentation, Sarah Coefield spoke about how public health air quality districts, emergency operations centers, and non-profit organizations are working toward the shared goal of creating cleaner indoor air opportunities for people. In Montana, where she lives and works, wildfires are a significant contributor to indoor fine particulate matter ($PM_{2.5}$), but so too is biomass burning for heating and cooking. In fact, Missoula County has a wood stove changeout and removal program that provides a bounty for homeowners to remove them and install a better source for heating and cooking that would dramatically improve both indoor and ambient air quality. Montana also has an asthma control program that sends program staff to a client's home to identify asthma triggers and provide a portable air cleaner with a HEPA filter to residents who have asthma triggers where they live.

The smoke generated during the extreme 2017 wildfire season drove home the inadequacy of standard wildfire smoke public health messaging, said Coefield, who referred specifically to the advice given at the time for people to stay inside their home. "If you have a smoke event that is a couple hours or a day, yes, if you go inside your home, close your doors and windows, you might be okay. When you have the same hazardous air pollution day after day for days, weeks, and months, that is no longer helpful," she said.

She noted that the role of public health as it relates to wildfire smoke has become much larger than it was a decade ago, and there are now three aspects to that role:

- communication: providing advice for creating clean air spaces and reducing exposure, providing health advisories and guidance documents, issuing smoke outlooks, and responding to public inquiries
- intervention: setting up cleaner-air shelters and providing portable air cleaners and N95 masks
- policy work: advocating for cleaner indoor air requirements, adopting updated ASHRAE guidance, canceling or postponing events, closing schools, and setting up safeguards for outdoor workers.

Regarding policy work, she said that interventions might work in a small community, but when millions of people are exposed to huge amounts of particulate matter, the need is for policies and actions beyond what the local health department can do to safeguard health.

Messaging about Wildfire Smoke

Missoula County takes a multipronged approach that starts with crafting smoke-ready messaging before the wildfire season begins and establishing a system for issuing daily wildfire smoke updates during the season. This includes setting up listservs, email distribution lists, websites, and blogs; crafting news releases; establishing a social media presence; preparing guidance documents, posters, and pamphlets; holding public meetings; and targeting messages to vulnerable groups. The website (www.montanawildfiresmoke.org), hosted and run by Climate Smart Missoula, serves as the main source of information on the internet, providing information on how to create clean indoor spaces and what to know about wildfire smoke. Since there are people who do not rely on the internet for information, her office has taken the thousands of words she has written about wildfire smoke and condensed them into posters for rural community grocery and hardware stores, for example, and pamphlets geared to specific audiences, such as older adults. The health department has also hosted smoke-ready workshops for business owners and building operations people so they can learn how to create cleaner indoor air spaces for their customers and employees.

These workshops, said Coefield, were adapted from the Smokewise Ashland program in Oregon.

She noted that Washington State has a Smoke-Ready Week during which multiple government agencies coordinate an effort to flood the airwaves with information about preparing for fire season before it starts. The goal is to get people to have their filters ready, make sure their HVAC system is in good shape, buy portable air cleaners, and make a plan before the smoke arrives.

Communication efforts do not stop once smoke does arrive. Coefield, as an air quality specialist in the health department, issues daily updates providing information about current air quality conditions, associated health advisories, what the fires are doing, where the smoke is coming from, where it is going to go, and how it is going to behave throughout the day. The idea, she said, it to give people information so they can plan their day. "I cannot make the smoke go away, but I can at least tell them what is happening, and that can give them some sort of agency and get them through the smoke event," said Coefield. She added that the updates also include mental health resources, because weeks and months of smoke can wear on people, as well as notices of areas with cleaner air in case people want to take a road trip to get some relief.

Interventions

Interventions during a smoke event include portable air cleaner distribution and instructions for creating a do-it-yourself air cleaner from a newer box fan and a good HVAC filter, which can be quite effective at removing particulate matter from indoor air. Coefield cautioned, however, that box fans built prior to 2011 had an unfortunate tendency to catch on fire, while newer fans have numerous safety features that prevent that from happening. She noted, too, that do-it-yourself air cleaners are louder than consumer-grade portable air cleaners. Interventions may also include evacuation (this happens more frequently in First Nation communities and in Canada).

In Missoula, her department has a cache of 125 portable air cleaners that she loans out to daycare centers and preschools to protect this vulnerable age group. She also keeps some in reserve for schools. When it comes to setting up an air cleaner distribution system, it is important to decide ahead of time how to select and triage recipients, whether the program will loan or give them away—she prefers giving them away—and if there will be follow-up to ensure that the recipients are using the cleaner correctly and replacing filters as needed. She noted that Ashland, Oregon, has a good system using text messages to keep track of the recipients, and she recommended the health department there as a good resource when setting up a distribution system. There is also the matter of how to fund the acquisition of a cache of air cleaners, since grant funding for this purpose is rare and usually competitive.

In recent years, said Coefield, interest in setting up cleaner air centers has grown, though this is a resource-intensive intervention. The least resource-intensive step is to preidentify publicly accessible buildings in the community that have advanced filtration and a plan for wildfire season; but she acknowledged that the odds of finding a building that meets such requirements before fire season are not high. An approach that has a moderate resource demand is to designate respite sites where people could go during business hours to get a few hours of relief. Those places, however, have to provide child care, snacks, entertainment, sanitation, and security. The most resource-intensive plan is to create 24-hour shelters, with food, sleeping facilities, and staff available.

Creating a cleaner indoor air center will likely require retrofitting the HVAC system to provide cleaner air and installation of sensors to monitor indoor particulate matter levels. It is also important to be prepared for the fact that smoke events can last for days, weeks, and even months, as happened in Missoula in 2017, and to consider who will use the space and if there is another, less expensive, and less resource-intensive intervention available. As with the air cleaner cache, there is also the need to secure funds to create a cleaner indoor air center. It may be less expensive, for example, to provide air cleaners to people.

The Missoula health department partnered with EPA in a study called Advance Science Partnerships for Indoor Reductions of Smoke Exposures[2] to better understand indoor air quality in publicly accessible buildings that people might go to as a respite from the smoke (Hagler et al., 2020). Coefield and her collaborators put low-cost sensors inside and outside a variety of buildings and found that nearly every building had MERV-8 filters (buildings that were supposed to have MERV-13 filters because they are LEED Platinum certified[3] did not. But the amount of particulate reduction in the buildings with MERV-8 filters varied significantly, as did different locations in the same building.

Coefield noted that the best reduction the study team found in the 2 years of field research was a 60 percent reduction in particulate matter— "not fantastic." "If there are 200 µg/m^3 outside, and you only have a 60

---

[2] Additional information is available at https://www.epa.gov/air-research/wildfire-study-advance-science-partnerships-indoor-reductions-smoke-exposures.

[3] LEED is a building certification program administered by the US Green Building Council that focuses on design, construction, and maintenance practices that promote environmental responsibility, indoor environmental quality, and energy conservation.

percent reduction, you still have unhealthy air quality inside the building," she said; for "preschools, schools, and day cares, you want better than that." She also reported that the project's engineer found many troubling issues with HVAC maintenance that explained a great deal of the results they found in the buildings.

Fortunately, said Coefield, some states are enacting policies to implement some of the actions necessary to protect the citizenry from extreme wildfire smoke events. Montana, for example, now has a rule that schools must limit infiltration during air pollution events and inspect their HVAC systems annually. Washington state has issued school closing guidance, and California has allocated funds for a Wildfire Smoke Clean Air Centers for Vulnerable Populations Incentive Pilot Program and will provide money to upgrade filtration systems and purchase portable air cleaners. California has also established worker protection rules that require companies that do not filter their air to provide respiratory protection during wildfire smoke events.

In addition, ASHRAE has published the framework for a proposed guideline (*Guideline 44P – Protecting Commercial Building Occupants During Wildfire Events*) that would advise building owners on how to make buildings with HVAC systems smoke ready (Javins et al., 2021). This guidance, said Coefield, should have more weight than anything she could say to a building owner.

In summary, Coefield said there is a need to know more about the most effective means of communicating to trigger behavior changes and to know what the most effective interventions are that will provide the biggest return on investment. Monitoring is needed, too, to show how effective cleaning technologies are during an extreme wildfire smoke event. And more work is needed at the policy level to create cleaner air spaces using existing technology that works.

DISCUSSION

Moderator Rengie Chan initiated the discussion by asking Batterman to elaborate on the findings of the field research he described in his talk. He reiterated some of the major parameters of that work and the need to address a number of personal and community-level factors—such as air filtration unit size, noise, and drafts—to encourage filter use. His Detroit-based study used trained personnel from the neighborhood to provide engagement and education to the participants on an ongoing basis. However, "while providing information on fine particulate matter exposure hazards is helpful," he said, "it's not the same as behavioral change." Study participants came from financially challenged households and spending

$400 to $500 a year for air filtration is not a viable option for people in those circumstances. At the same time, Batterman noted that many people with sufficient means still buy the cheapest filters available, so there are opportunities to improve labeling and merchandising practices to encourage different purchasing choices. He acknowledged that even he does not completely understand the current labeling system for filters he finds at the hardware store.

Graham responded to a question about the role of sociocultural factors and background in the selection of participants in her study by noting that they were all in the San Francisco Bay Area, where wildfire events are a regular occurrence and most individuals are instructed at some point to pay attention to the air quality around them. She acknowledged that culture can also influence perceptions and decision making, and said that it would be interesting to replicate her study in another location where air quality awareness was different to better examine such questions. She and her colleagues had asked participants whether they had lived in locations outside the Bay Area to begin to tease this out.

Chan asked Coefield about challenges in helping schools to interpret the air quality data they get from sensors installed in their buildings. Currently, said Coefield, there is no requirement for schools to monitor indoor particulate matter levels or to provide cleaner indoor air for their students. "We need to get to a place where we actually require cleaner indoor air so we can see the types of changes that we need, because if everything is voluntary, folks find other uses for their money," she said.

Asked about the value of adding ultraviolet, ionization, and photocatalytic oxidation systems to existing HVAC systems, both to remove indoor particulate matter and to address COVID-19 concerns, Siegel called this the "kitchen sink" approach to air cleaners and said that none of those devices have been demonstrated to add any value. In some cases, there may even be negative consequences. "That does not mean that those technologies maybe do not have a place, just that we lack evidence of efficacy, and I think because of the potential harm of some of the devices and some of the technologies out there,…we have to be careful," said Siegel. Batterman reported that some city and state governments believe that plasma air purifier units work to limit exposures, and he is finding it challenging to convince them otherwise.

Singer reiterated an earlier message that basic filtration works well as long as the system is moving air through the filter. One of the nice things about portable air filters, he said, is that many come with a certified Association of Home Appliance Manufacturers rating that tells the consumer that if they operate it at the set speed, they will know what they are getting and not have to worry about bypass—and many of them are more efficient to operate than a central HVAC system.

Richard Corsi remarked that a good, portable HEPA filtration device installed in a typical classroom of 25 students costs about $10 per student per year, compared to the $15,000 annual cost of educating that student. He would like discussion of the health benefits of limiting indoor particulate matter exposures to focus on that cost per student and believes it would get more attention. Singer said that was an excellent point, particularly given that classroom HEPA filtration units would be one of the easier solutions for a school district, which could do a mass procurement and not need to upgrade an HVAC system that might be 30 to 50 years old. "It's quick, it's affordable, and the maintenance is very simple and we all need to keep hammering away that sometimes the simple solution is the best solution."

A participant asked if higher-rated MERV filters can create problems in older HVAC systems, and Siegel replied that low air flow can be a problem in older central, forced-air systems. He countered, though, that the filter is rarely the problem; instead, often there is not enough airflow through the negative pressure return side of the system. The real answer, he said, is to fix the fundamental problem with the system and then put in a good filter. He also noted that the common assumption that a better, more efficient filter produces a higher pressure drop is not borne out by the evidence. He recommended replacing the filter slot to address bypass with a gasketed filter slot that can accommodate a 2-inch or 4-inch filter, which will have a lower pressure drop and provide the needed performance. The key is to educate consumers and installers. Batterman added that many buildings still don't have air conditioning: it is difficult to provide adequate filtration when windows are open.

Chan asked the speakers to comment on the idea of focusing more on reducing infiltration or achieving tighter building envelopes as a way to mitigate indoor exposure to outdoor particulate matter. Batterman replied that his group has looked at dozens of schools and found chronic problems with lack of ventilation or inadequate ventilation for the capacity of the space, which has to be addressed. Coefield added that tightening the building envelope can be a good idea for homeowners in areas with poor outdoor air quality, but the trade-off is that indoor-generated particles are trapped inside and it's necessary to mitigate that exposure. Graham noted the importance of educating the public so that people understand how their behaviors and their daily actions influence their indoor air quality. "That is central to all of these issues that we are talking about because that is going to dictate the success or not of how we continue to engage within those spaces regardless of the building intervention," said Graham.

A question was then raised about the effectiveness of filtration on particulate matter with different chemical compositions. Gall replied that the primary determinant of removal efficiency was particle diameter, which means that a filter can be effective for particulates originating from multiple sources in the same size range. Singer qualified, though, that organic compounds can accumulate on filters during events like wildfires and lead to both odors and subsequent exposures. Filter replacement isn't cheap and there is a need to refine strategies for determining when replacement is appropriate and whether there are other strategies for mitigating this problem.

10

Workshop Summary and Closing Reflections

To conclude the workshop, Planning Committee Chair Richard Corsi summarized the key points he heard during the 3 days of presentations and discussions. One point that came across was that great technology is worthless if it is not used, and poor technology is worthless even if it is used, highlighting the importance of both technology and human behavior and decision making. He then noted that indoor environments are a complex system, starting with the fact that indoor fine particulate matter ($PM_{2.5}$) of outdoor origin mixes with $PM_{2.5}$ generated indoors and chemical transformations occur indoors. Ventilation is an important defining factor, as is deposition of particles on surfaces and engineered controls. It is therefore important to understand the exposure of occupants, the inhalation dose, where particles go in the respiratory system, and what the health effects are.

Contributing to the complexity of the system are significant temporal variations for sources and indoor concentrations of $PM_{2.5}$, as well as significant spatial gradients, both outdoors and indoors. There are also differences in how occupants perceive indoor air quality and how they behave and respond, as well as different policies that affect how indoor $PM_{2.5}$ is handled. Superimposed on all of this are socioeconomic issues that further complicate these matters. "This underscores the fact that we need an interdisciplinary approach to deal with these problems," said Corsi.

OUTDOOR $PM_{2.5}$

Starting with the presentations on outdoor sources of $PM_{2.5}$, Corsi listed the following highlights:

- There are significant temporal and spatial gradients with respect to outdoor sources of $PM_{2.5}$ that affect indoor $PM_{2.5}$, and there are significant environmental justice issues associated with the locations of underserved communities with respect to many of these sources.
- Outdoor-to-indoor transport of $PM_{2.5}$ across the building envelope results in an average of half of the outdoor $PM_{2.5}$ particles infiltrating to indoors, although there are many factors that affect that infiltration and large variation in how much of it occurs.
- Analysis of microenvironmental $PM_{2.5}$ exposures suggests that about 70 percent of $PM_{2.5}$ exposure occurs in the home and about 20 percent in other buildings, which means that less than 10 percent of an individual's exposure occurs outdoors.
- Roughly half of the exposure indoors is to $PM_{2.5}$ of outdoor origin and half from $PM_{2.5}$ of indoor origin.
- Indoor sources of $PM_{2.5}$ include combustion (cooking, candles, and fuels), food being cooked, phase change as cooking oils heat and cool, mechanical (ultrasonic humidifiers and vacuuming), biological (respiratory aerosols and mold spores), and chemical reactions that generate secondary organic aerosols.
- Emissions may be high during cooking, with particle numbers dominated by the fuel source. Particulate emission levels and composition are highly variable, and there is little information about what this means for health.
- Transformations occur as outdoor $PM_{2.5}$ infiltrates indoors, where those particles encounter larger surface-to-volume ratios and fewer oxidants. Particles can shrink through evaporation and they can adsorb volatile and semivolatile organic compounds that are present at much higher concentrations indoors. There is also thermal partitioning that occurs indoors. The question is whether these transformations have relevance for human health.
- Relative humidity affects the water content of $PM_{2.5}$, which can change the size of the particles as well as the chemistry and microbiology in and on the particles. It is not known what the optimum environmental conditions are to reduce the health effects of $PM_{2.5}$.

HEALTH EFFECTS OF INDOOR $PM_{2.5}$

Corsi reviewed the following highlights of the presentations on the health effects of indoor $PM_{2.5}$:

- The vast burden of morbidity from air pollution is due to increased cardiovascular disease, largely because cardiovascular disease is highly prevalent in adults living in the developed world.
- Given that $PM_{2.5}$ of outdoor origin accounts for about 70 percent of the total dose, the epidemiologic findings on outdoor $PM_{2.5}$ apply to indoor exposures.
- Increases in indoor $PM_{2.5}$ are associated with increases in asthma and chronic obstructive pulmonary disease morbidity, but portable HEPA filtration can reduce symptoms.
- There is some evidence from animal studies that $PM_{2.5}$ of indoor origin produces more inflammation and cytotoxicity than $PM_{2.5}$ of outdoor origin.
- Assessing indoor $PM_{2.5}$ is complex, and personal measurements are best when possible, though they are cumbersome. Low-cost consumer instruments for $PM_{2.5}$ are making measurements more common, but protocols and reference standards are missing, and the accuracy and performance of these devices in real-life settings are not impressive.
- Multiple dimensions of complexity are associated with indoor $PM_{2.5}$, including their broad size range, chemical composition, high temporal variability for particle concentrations, and high spatial variability. Since no instrument can measure all of these factors, it is important to decide which measurements are important, where they should be made, and when to make them. The answers depend in part on whether the concern is short-term or long-term health effects.
- Inhalation dose and location of where particles deposit in the respiratory system are both important factors in determining possible health effects.
- Moving forward with the use of low-cost sensors will require understanding that individual perception of indoor air quality is important, and evaluating whether sensors can help with those perceptions.
- People want information that provides guidance on the actions they should take or not take. Clear, graphical, and actionable information is important to encourage desired occupant behavior.
- Improved filtration has been associated with subclinical cardiovascular benefits, increased birth weight, reduced asthma morbidity,

and higher test scores in schools. Portable HEPA cleaners have been associated with substantial reductions in $PM_{2.5}$ exposure and an increase in symptom-free days for children with asthma. These studies are limited in number, and there is a need to do better at communicating the benefits of improved filtration and develop standards to help individuals realize those benefits.

MITIGATION OF INDOOR $PM_{2.5}$

Moving on to the presentations on approaches to mitigating indoor $PM_{2.5}$, Corsi listed the following highlights:

- Local exhaust of emissions from cooking, a major source of indoor $PM_{2.5}$, can be achieved with a range hood, downdraft exhaust system, ceiling or wall exhaust, or open window.
- Hood capture efficiency is important, and using back burners rather than front burners results in better removal of $PM_{2.5}$. Regarding hoods, recirculating systems fare poorly in mitigating exposure compared to those that exhaust to the outdoors. Communicating this to the public is important.
- Occupant behavior is key given that most people use their range hoods less than half the time when cooking, in part because they are noisy. Are there approaches such as sensors that either alert occupants to turn on their range hoods or automatic systems that can improve this situation?
- Filtration works only when there is a good filter that is properly installed to minimize bypass, when the filtration system is running and producing enough air flow, and when the filter is replaced regularly. Other air cleaning technologies are not well defined, lack independent evaluation, and may even be harmful.
- Given that some 15 percent of US schools are within 250 meters of a major roadway, it is important to mitigate indoor $PM_{2.5}$ levels in schools. Traffic-related air pollution affects not only respiratory health conditions such as asthma but also cognition. This is an environmental justice issue because schools with higher percentages of Hispanic, Black, and Asian students are more likely to be located near such outdoor PM sources.
- Significant reductions in traffic-related air pollution can be achieved with the proper filters and an efficiently operating HVAC system. Portable HEPA filtration systems installed in classrooms can be an effective and cost-effective exposure mitigation strategy for schools.

- Portable HEPA cleaners in the home can produce significant reductions in indoor $PM_{2.5}$ and may thus reduce the frequency of asthma symptoms. This benefit can only accrue, though, if the devices are used, which may not happen on a regular basis because units may produce drafts and noise, and the cost of operating these devices can be prohibitive for families of more limited means.
- Messaging is critical for reducing exposures to $PM_{2.5}$ during wildfires, as is preparation before an event occurs. Portable HEPA cleaners can be valuable mitigation tools during these and other high outdoor particulate matter events.
- There are substantial exposure and health disparities related to socioeconomic status, with underserved communities carrying a high burden due to proximity to outdoor sources, higher occupant density, older ventilation and exhaust systems, and economic challenges that affect the use of practical mitigation approaches. Other factors, like the amount of time spent in the home and cooking practices, also affect exposure.

Corsi concluded the session by acknowledging the US Environmental Protection Agency's sponsorship of the workshop and thanking the presenters and the planning committee. The workshop was then adjourned.

References

Adamkiewicz G, Zota AR, Fabian MP, Chahine T, Julien R, Spengler JD, Levy JI. 2011. Moving environmental justice indoors: Understanding structural influences on residential exposure patterns in low-income communities. *American Journal of Public Health* 101(S1):S238–45.

Alavy M, Li T, Siegel JA. 2020. Energy use in residential buildings: Analyses of high-efficiency filters and HVAC fans. *Energy and Buildings* 209:109697.

Alavy M, Siegel JA. 2019. IAQ and energy implications of high efficiency filters in residential buildings: A review (RP-1649). *Science and Technology for the Built Environment* 25(3):261–71.

Allen RW, Barn P. 2020. Individual- and household-level interventions to reduce air pollution exposures and health risks: A review of the recent literature. *Current Environmental Health Reports* 7(4):424–40.

Alves CA, Vicente ED, Evtyugina M, Vicente AMP, Sainnokhoi T-A, Kováts N. 2021. Cooking activities in a domestic kitchen: Chemical and toxicological profiling of emissions. *Science of the Total Environment* 772:145412.

Avery AM, Waring MS, DeCarlo PF. 2019a. Human occupant contribution to secondary aerosol mass in the indoor environment. *Environmental Science: Processes & Impacts* 21(8):1301–12.

Avery AM, Waring MS, DeCarlo PF. 2019b. Seasonal variation in aerosol composition and concentration upon transport from the outdoor to indoor environment. *Environmental Science: Processes & Impacts* 21(3):528–47.

Azimi P, Stephens B. 2020. A framework for estimating the US mortality burden of fine particulate matter exposure attributable to indoor and outdoor microenvironments. *Journal of Exposure Science & Environmental Epidemiology* 30:271–84.

Barbosa S, Bermudez R, Cassmassi J, Cheung K, Eden R, Lee S-M, Low J, Ospital J, Pakbin P, Polidori A, and 3 others. 2015. *Multiple Air Toxics Exposure Study in the South Coast Air Basin (MATES IV)*. Diamond Bar, CA: South Coast Air Quality Management District.

Barn P, Larson T, Noullett M, Kennedy S, Copes R, Brauer M. 2008. Infiltration of forest fire and residential wood smoke: An evaluation of air cleaner effectiveness. *Journal of Exposure Science & Environmental Epidemiology* 18(5):503–11.

Barn P, Gombojav E, Ochir C, Boldbaatar B, Beejin B, Naidan G, Galsuren J, Legtseg B, Byambaa T, Hutcheon JA, and 8 others. 2018. The effect of portable HEPA filter air cleaner use during pregnancy on fetal growth: The UGAAR randomized controlled trial. *Environment International* 121:981–89.

Barry AC, Mannino DM, Hopenhayn C, Bush H. 2010. Exposure to indoor biomass fuel pollutants and asthma prevalence in southeastern Kentucky: Results from the Burden of Lung Disease (BOLD) study. *Journal of Asthma* 47(7):735–41.

Bates JT, Fang T, Verma V, Zeng L, Weber RJ, Tolbert PE, Abrams JY, Sarnat SE, Klein M, Mulholland JA, Russell AG. 2019. Review of acellular assays of ambient particulate matter oxidative potential: methods and relationships with composition, sources, and health effects. *Environmental Science & Technology* 53(8):4003–19.

Batterman S, Du L, Mentz G, Mukherjee B, Parker E, Godwin C, Chin JY, O'Toole A, Robins T, Rowe Z, Lewis T. 2012. Particulate matter concentrations in residences: An intervention study evaluating stand-alone filters and air conditioners. *Indoor Air* 22(3):235–52.

Batterman S, Du L, Parker E, Robins T, Lewis T, Mukherjee B, Ramirez E, Rowe Z, Brakefield-Caldwell W. 2013. Use of free-standing filters in an asthma intervention study. *Air Quality, Atmosphere & Health* 6(4):759–67.

Batterman S, Marval J, Tronville P, Goodwin C. 2021. Comparison and performance of standard and high-performance filters in laboratory and school environments. Paper presented at *RoomVent 2020*, March 1–15, Torino, Italy.

Bi J, Wildani A, Chang HH, Liu Y. 2020. Incorporating low-cost sensor measurements into high-resolution $PM_{2.5}$ modeling at a large spatial scale. *Environmental Science & Technology* 54(4):2152–62.

Bi J, Wallace LA, Sarnat JA, Liu Y. 2021. Characterizing outdoor infiltration and indoor contribution of $PM_{2.5}$ with citizen-based low-cost monitoring data. *Environmental Pollution* 276:116763.

Brehmer C, Norris C, Barkjohn KK, Bergin MH, Zhang J, Cui X, Teng Y, Zhang Y, Black M, Li Z, and 2 others. 2020. The impact of household air cleaners on the oxidative potential of $PM_{2.5}$ and the role of metals and sources associated with indoor and outdoor exposure. *Environmental Research* 181:108919.

Breysse PN, Diette GB, Matsui EC, Butz AM, Hansel NN, McCormack MC. 2010. Indoor air pollution and asthma in children. *Proceedings of the American Thoracic Society* 7(2):102–106.

Breysse J, Jacobs DE, Weber W, Dixon S, Kawecki C, Aceti S, Lopez J. 2011. Health outcomes and green renovation of affordable housing. *Public Health Reports* 126(1 Suppl.):64–75.

Breysse J, Dixon S, Gregory J, Philby M, Jacobs DE, Krieger J. 2014. Effect of weatherization combined with community health worker in-home education on asthma control. *American Journal of Public Health* 104(1):e57–e64.

Brook RD. 2008. Cardiovascular effects of air pollution. *Clinical Science (London)* 115(6):175–87.

Butz AM, Matsui EC, Breysse P, Curtin-Brosnan J, Eggleston P, Diette G, Williams D, Yuan J, Bernert JT, Rand C. 2011. A randomized trial of air cleaners and a health coach to improve indoor air quality for inner-city children with asthma and secondhand smoke exposure. *Archives of Pediatrics and Adolescent Medicine* 165(8):741–48.

Carslaw N, Mota T, Jenkin ME, Barley MH, McFiggans G. 2012. A significant role for nitrate and peroxide groups on indoor secondary organic aerosol. *Environmental Science & Technology* 46(17):9290–98.

Chafe, ZA, Brauer M, Klimont Z, Dingenen RV, Mehta S, Rao S, Riahi K, Dentener F, Smith KR. 2014. Household cooking with solid fuels contributes to ambient $PM_{2.5}$ air pollution and the burden of disease. *Environmental Health Perspectives* 122(12):1314–20.

Chen C, Zhao B. 2011. Review of relationship between indoor and outdoor particles: I/O ratio, infiltration factor and penetration factor. *Atmospheric Environment* 45(2):275–88.

Chen TM, Gokhale J, Shofer S, Kuschner WG. 2007. Outdoor air pollution: Particulate matter health effects. *American Journal of the Medical Sciences* 333(4):235–43.

Chen WR, Kim Y-S, Less BD, Singer BC, Walker IS. 2020. *Ventilation and Air Quality in New California Homes with Gas Appliances and Mechanical Ventilation.* LBNL–2001200R1. Berkeley, CA: Lawrence Berkeley National Laboratory.

Chojnowski DB, Nigro PJ, Siegel JA, Kosar DR. 2009. Laboratory measurements to quantify the effect of bypass on filtration efficiency. *ASHRAE Transactions* 115(1):199–207.

Coleman B, Lunden M, Destaillats H, Nazaroff W. 2008. Secondary organic aerosol from ozone-initiated reactions with terpene-rich household products. *Atmospheric Environment* 42:8234–45.

Collins DB, Wang C, Abbatt JPD. 2018. Selective uptake of third-hand tobacco smoke components to inorganic and organic aerosol particles. *Environmental Science & Technology* 52(22):13195–201.

Cummings BE, Waring MS. 2019. Predicting the importance of oxidative aging on indoor organic aerosol concentrations using the two-dimensional volatility basis set (2D-VBS). *Indoor Air* 29(4):616–29.

Cummings BE, Li Y, DeCarlo PF, Shiraiwa M, Waring MS. 2020. Indoor aerosol water content and phase state in US residences: Impacts of relative humidity, aerosol mass and composition, and mechanical system operation. *Environmental Science: Processes & Impacts* 22(10):2031–57.

Daellenbach KR, Uzu G, Jiang J, Cassagnes L-E, Leni Z, Vlachou A, Stefenelli G, Canonaco F, Weber S, Segers A, and 11 others. 2020. Sources of particulate-matter air pollution and its oxidative potential in Europe. *Nature* 587(7834):414–19.

DeCarlo PF, Avery AM, Waring MS. 2018. Thirdhand smoke uptake to aerosol particles in the indoor environment. *Science Advances* 4(5):eaap8368.

Demanega I, Mujan I, Singer BC, An elkovi AS, Babich F, Licina D. 2021. Performance assessment of low-cost environmental monitors and single sensors under variable indoor air quality and thermal conditions. *Building and Environment* 187:107415.

Destaillats H, Lunden MM, Singer BC, Coleman BK, Hodgson AT, Weschler CJ, Nazaroff WW. 2006. Indoor secondary pollutants from household product emissions in the presence of ozone: A bench-scale chamber study. *Environmental Science & Technology* 40(14):4421–28.

Di Q, Kloog I, Koutrakis P, Lyapustin A, Wang Y, Schwartz J. 2016. Assessing $PM_{2.5}$ exposures with high spatiotemporal resolution across the continental United States. *Environmental Science & Technology* 50(9):4712–21.

Di Q, Wang Y, Zanobetti A, Wang Y, Koutrakis P, Choirat C, Dominici F, Schwartz JD. 2017. Air pollution and mortality in the Medicare population. *New England Journal of Medicine* 376(26):2513–22.

Do K, Yu H, Velasquez J, Grell-Brisk M, Smith H, Ivey CE. 2021. A data-driven approach for characterizing community scale air pollution exposure disparities in inland southern California. *Journal of Aerosol Science* 152:105704.

Doiron D, de Hoogh K, Probst-Hensch N, Fortier I, Cai Y, De Matteis S, Hansell AL. 2019. Air pollution, lung function and COPD: Results from the population-based UK Biobank study. *European Respiratory Journal* 54(1):1802140.

Du L, Batterman S, Parker E, Godwin C, Chin JY, O'Toole A, Robins T, Brakefield-Caldwell W, Lewis T. 2011. Particle concentrations and effectiveness of free-standing air filters in bedrooms of children with asthma in Detroit, Michigan. *Building and Environment* 46(11):2303–13.

Eichler CMA, Hubal EAC, Xu Y, Cao J, Bi C, Weschler CJ, Salthammer T, Morrison GC, Koivisto AJ, Zhang Y, and 16 others. 2021. Assessing human exposure to SVOCs in materials, products, and articles: A modular mechanistic framework. *Environmental Science & Technology* 55(1):25–43.

Emerson, EW, Hodshire AL, DeBolt HM, Bilsback KR, Pierce JR, McMeeking GR, Farmer DK. 2020. Revisiting particle dry deposition and its role in radiative effect estimates. *Proceedings of the National Academy of Sciences* 117(42):26076–82.

Eriksson AC, Andersen C, Krais AM, Nøjgaard JK, Clausen P-A, Gudmundsson A, Wierzbicka A, Pagels J. 2020. Influence of airborne particles' chemical composition on SVOC uptake from PVC flooring: Time-resolved analysis with aerosol mass spectrometry. *Environmental Science & Technology* 54(1):85–91.

Fabian MP, Adamkiewicz G, Stout NK, Sandel M, Levy JI. 2014. A simulation model of building intervention impacts on indoor environmental quality, pediatric asthma, and costs. *Journal of Allergy and Clinical Immunology* 133:77–84.

Farmer DK, Vance ME, Abbatt JPD, Abeleira A, Alves MR, Arata C, Boedicker E, Bourne S, Cardoso-Saldaña F, Corsi R, and 20 others. 2019. Overview of HOMEChem: House Observations of Microbial and Environmental Chemistry. *Environmental Science: Processes & Impacts* 21(8):1280–300.

Fisk WJ, Chan WR. 2017. Health benefits and costs of filtration interventions that reduce indoor exposure to $PM_{2.5}$ during wildfires. *Indoor Air* 27(1):191–204.

Fisk WJ, Lei-Gomes Q, Mendell MJ. 2007. Meta analyses of the associations of respiratory health effects with dampness and mold in homes. *Indoor Air* 17(4):284–296.

Freedman MA, Ott E-JE, Marak KE. 2019. Role of pH in aerosol processes and measurement challenges. *Journal of Physical Chemistry A* 123(7):1275–84.

Gaffron P, Niemeier D. 2015. School locations and traffic emissions: Environmental (in)justice findings using a new screening method. *International Journal of Environmental Research and Public Health* 12(2):2009–25.

Gao X, Xu Y, Cai Y, Shi J, Chen F, Lin Z, Chen T, Xia Y, Shi W, Zhao Z. 2019. Effects of filtered fresh air ventilation on classroom indoor air and biomarkers in saliva and nasal samples: A randomized crossover intervention study in preschool children. *Environmental Research* 179:108749.

Garofalo LA, Pothier MA, Levin EJT, Campos T, Kreidenweis SM, Farmer DK. 2019. Emission and evolution of submicron organic aerosol in smoke from wildfires in the western United States. *ACS Earth and Space Chemistry* 3(7):1237–47.

Gauderman WJ, Urman R, Avol E, Berhane K, McConnell R, Rappaport E, Chang R, Lurmann F, Gilliland F. 2015. Association of improved air quality with lung development in children. *New England Journal of Medicine* 372(10):905–13.

GBD 2016 Risk Factors Collaborators. 2017. Global, regional, and national comparative risk assessment of 84 behavioural, environmental and occupational, and metabolic risks or clusters of risks, 1990–2016: A systematic analysis for the Global Burden of Disease study 2016. *The Lancet* 390(10100):1345–422.

Gelfand AE, Zhu L, Carlin BP. 2001. On the change of support problem for spatio-temporal data. *Biostatistics* 2(1):31–45.

Gold DR, Adamkiewicz G, Arshad SH, Celedón JC, Chapman MD, Chew GL, Cook DN, Custovic A, Gehring U, Gern JE, and 21 others. 2017. NIAID, NIEHS, NHLBI, and MCAN workshop report: The indoor environment and childhood asthma-implications for home environmental intervention in asthma prevention and management. *Journal of Allergy and Clinical Immunology* 140(4):933–49.

Goldstein AH, Nazaroff WW, Weschler CJ, Williams J. 2021. How do indoor environments affect air pollution exposure? *Environmental Science & Technology* 55(1):100–108.

Graham LT, Parkinson T, Schiavon S. 2021. Lessons learned from 20 years of CBE's occupant surveys. *Buildings and Cities* 2(1):166–84.
Grineski SE, Collins TW. 2018. Geographic and social disparities in exposure to air neurotoxicants at US public schools. *Environmental Research* 161:580–87.
Guo C, Zhang Z, Lau AKH, Lin CQ, Chuang YC, Chan J, Jiang WK, Tam T, Yeoh EK, Chan TC, and 2 others. 2018. Effect of long-term exposure to fine particulate matter on lung function decline and risk of chronic obstructive pulmonary disease in Taiwan: A longitudinal, cohort study. *The Lancet Planetary Health* 2(3):e114–25.
Haaland D, Siegel JA. 2017. Quantitative filter forensics for indoor particle sampling. *Indoor Air* 27(2):364–76.
Hagler G, Holder A, Katz S, Robarge G, Hassett-Sipple B, Cascio W, Coefield S, Schmidt B, Neale G, Noonan C, Weiler E. 2020. Field study summary: $PM_{2.5}$ measurements inside and outside of buildings in Missoula, MT during summer 2019. EPA webinar presentation, February 20, Chapel Hill, NC.
Hanley JT, Ensor DS, Smith DD, Sparks LE. 1994. Fractional aerosol filtration efficiency of in-duct ventilation air cleaners. *Indoor Air* 4(3):169–78.
Hansel NN, McCormack MC, Belli AJ, Matsui EC, Peng RD, Aloe C, Paulin L, Williams DA, Diette GB, Breysse PN. 2013. In-home air pollution is linked to respiratory morbidity in former smokers with chronic obstructive pulmonary disease. *American Journal of Respiratory and Critical Care Medicine* 187(10):1085–90.
Hansel NN, Putcha N, Woo H, Peng R, Diette GB, Fawzy A, Wise RA, Romero K, Davis MF, Rule AM, and 4 others. 2021. Randomized clinical trial of air cleaners to improve indoor air quality and COPD health: Results of the CLEAN AIR STUDY. Preprints with *The Lancet*, posted February 16, 2021.
Happo MS, Sippula O, Jalava PI, Rintala H, Leskinen A, Komppula M, Kuuspalo K, Mikkonen S, Lehtinen K, Jokiniemi J, Hirvonen MR. 2014. Role of microbial and chemical composition in toxicological properties of indoor and outdoor air particulate matter. *Particle and Fibre Toxicology* 11:60.
He L-Y, Hu M, Huang X-F, Yu B-D, Zhang Y-H, Liu D-Q. 2004. Measurement of emissions of fine particulate organic matter from Chinese cooking. *Atmospheric Environment* 38(38):6557–64.
HEI Panel on the Health Effects of Traffic-Related Air Pollution. 2010. *Traffic-Related Air Pollution: A Critical Review of the Literature on Emissions, Exposure, and Health Effects*. Boston, MA: Health Effects Institute.
Henderson DE, Milford JB, Miller SL. 2005. Prescribed burns and wildfires in Colorado: Impacts of mitigation measures on indoor air particulate matter. *Journal of the Air & Waste Management Association* 55(10):1516–26.
Holstius DM, Pillarisetti A, Smith KR, Seto E. 2014. Field calibrations of a low-cost aerosol sensor at a regulatory monitoring site in California. *Atmospheric Measurement Techniques* 7(4):1121–31.
Horak F, Studnicka M, Gartner C, Spengler JD, Tauber E, Urbanek R, Veiter A, Frischer T. 2002. Particulate matter and lung function growth in children: A 3-yr follow-up study in Austrian schoolchildren. *European Respiratory Journal* 19(5):838.
Huang W, Wang G, Lu SE, Kipen H, Wang Y, Hu M, Lin W, Rich D, Ohman-Strickland P, Diehl SR, and 5 others. 2012. Inflammatory and oxidative stress responses of healthy young adults to changes in air quality during the Beijing Olympics. *American Journal of Respiratory and Critical Care Medicine* 186(11):1150–59.
Iaccarino L, Joie RL, Lesman-Segev OH, Lee E, Hanna L, Allen IE, Hillner BE, Siegel BA, Whitmer RA, Carrillo MC, and 2 others. 2021. Association between ambient air pollution and amyloid positron emission tomography positivity in older adults with cognitive impairment. *JAMA Neurology* 78(2):197–207.

Ivey CE, Holmes HA, Hu YT, Mulholland JA, Russell AG. 2015. Development of $PM_{2.5}$ source impact spatial fields using a hybrid source apportionment air quality model. *Geoscientific Model Development* 8(7):2153–65.

Ivey CE, Holmes HA, Hu Y, Mulholland JA, Russell AG. 2016. A method for quantifying bias in modeled concentrations and source impacts for secondary particulate matter. *Frontiers of Environmental Science & Engineering* 10(5):14.

Ivey CE, Balachandran S, Colgan S, Hu Y, Holmes HA. 2019. Investigating fine particulate matter sources in Salt Lake City during persistent cold air pool events. *Atmospheric Environment* 213:568–78.

Jaffe DA, O'Neill SM, Larkin NK, Holder AL, Peterson DL, Halofsky JE, Rappold AG. 2020. Wildfire and prescribed burning impacts on air quality in the United States. *Journal of the Air & Waste Management Association* 70(6):583–615.

Janssen NA, Hoek G, Simic-Lawson M, Fischer P, Van Bree, L, Ten Brink H, Keuken M, Atkinson RW, Anderson HR, Brunekreef B, Cassee, FR. 2011. Black carbon as an additional indicator of the adverse health effects of airborne particles compared with PM_{10} and $PM_{2.5}$. *Environmental Health Perspectives* 119(12):1691–1699.

Javins T, Robarge G, Snyder EG, Nilsson G, Emmerich SJ. 2021. Protecting building occupants from smoke during wildfire and prescribed burn events. *ASHRAE Journal* 63(3):38–40, 42–43.

Jayaratne R, Liu X, Ahn K-H, Asumadu-Sakyi A, Fisher G, Gao J, Mabon A, Mazaheri M, Mullins B, Nyaku M, and 5 others. 2020. Low-cost $PM_{2.5}$ sensors: An assessment of their suitability for various applications. *Aerosol and Air Quality Research* 20(3):520–32.

Jhun I, Gaffin JM, Coull BA, Huffaker MF, Petty CR, Sheehan WJ, Baxi SN, Lai PS, Kang CM, Wolfson JM, and 3 others. 2017. School environmental intervention to reduce particulate pollutant exposures for children with asthma. *Journal of Allergy and Clinical Immunology: In Practice* 5(1):154–59 e3.

Johnson AM, Waring MS, DeCarlo PF. 2017. Real-time transformation of outdoor aerosol components upon transport indoors measured with aerosol mass spectrometry. *Indoor Air* 27(1):230–40.

Jovašević-Stojanović M, Bartonova A, Topalović D, Lazović I, Pokrić B, Ristovski Z. 2015. On the use of small and cheaper sensors and devices for indicative citizen-based monitoring of respirable particulate matter. *Environmental Pollution* 206:696–704.

Karner AA, Eisinger DS, Niemeier DA. 2010. Near-roadway air quality: Synthesizing the findings from real-world data. *Environmental Science & Technology* 44(14):5334–44.

Katz EF, Guo H, Campuzano-Jost P, Day DA, Brown WL, Boedicker E, Pothier M, Lunderberg DM, Patel S, Patel K, and 8 others. 2021. Quantification of cooking organic aerosol in the indoor environment using Aerodyne aerosol mass spectrometers. *Aerosol Science and Technology* 55(10):1099–114.

Kearney J, Wallace L, MacNeill M, Héroux M-E, Kindzierski W, Wheeler A. 2014. Residential infiltration of fine and ultrafine particles in Edmonton. *Atmospheric Environment* 94:793–805.

Kennedy C. 2021. *Risk of Very Large Fires Could Increase Sixfold by Mid-century in the US*. https://www.climate.gov/news-features/featured-images/risk-very-large-fires-could-increase-sixfold-mid-century-us, Published August 26, 2015, updated July 2, 2021. Silver Spring, MD: National Oceanic and Atmospheric Administration.

Khurshid SS, Emmerich S, Persily A. 2019. Oxidative potential of particles at a research house: Influencing factors and comparison with outdoor particles. *Building and Environment* 163:106275.

Kingsley SL, Eliot MN, Carlson L, Finn J, MacIntosh DL, Suh HH, Wellenius GA. 2014. Proximity of US schools to major roadways: A nationwide assessment. *Journal of Exposure Science & Environmental Epidemiology* 24(3):253–59.

Klein F, Baltensperger U, Prévôt ASH, El Haddad I. 2019. Quantification of the impact of cooking processes on indoor concentrations of volatile organic species and primary and secondary organic aerosols. *Indoor Air* 29(6):926–42.

Klepeis NE, Nelson WC, Ott WR, Robinson JP, Tsang AM, Switzer P, Behar JV, Hern SC, Engelmann WH. 2001. The National Human Activity Pattern Survey (NHAPS): A resource for assessing exposure to environmental pollutants. *Journal of Exposure Science & Environmental Epidemiology* 11(3):231–52.

Koehler KA, Clark P, Volckens J. 2009. Development of a sampler for total aerosol deposition in the human respiratory tract. *Annals of Occupational Hygiene* 53(7):731–38.

Koehler K, Good N, Wilson A, Mölter A, Moore BF, Carpenter T, Peel JL, Volckens J. 2019. The Fort Collins commuter study: Variability in personal exposure to air pollutants by microenvironment. *Indoor Air* 29(2):231–41.

Kozawa KH, Fruin SA, Winer AM. 2009. Near-road air pollution impacts of goods movement in communities adjacent to the ports of Los Angeles and Long Beach. *Atmospheric Environment* 43(18):2960–70.

Kristensen K, Lunderberg DM, Liu Y, Misztal PK, Tian Y, Arata C, Nazaroff WW, Goldstein AH. 2019. Sources and dynamics of semivolatile organic compounds in a single-family residence in northern California. *Indoor Air* 29(4):645–55.

Kroll J, Seinfeld J. 2008. Chemistry of secondary organic aerosol: Formation and evolution of low-volatility organics in the atmosphere. *Atmospheric Environment* 42(16):3593–24.

Kruza M, McFiggans G, Waring MS, Wells JR, Carslaw N. 2020. Indoor secondary organic aerosols: Towards an improved representation of their formation and composition in models. *Atmospheric Environment* 240:117784.

Kumar P, Morawska L, Martani C, Biskos G, Neophytou M, Di Sabatino S, Bell M, Norford L, Britter R. 2015. The rise of low-cost sensing for managing air pollution in cities. *Environment International* 75:199–205.

Kunkel SA, Azimi P, Zhao H, Stark BC, Stephens B. 2017. Quantifying the size-resolved dynamics of indoor bioaerosol transport and control. *Indoor Air* 27(5):977–87.

Laguerre A, George LA, Gall ET. 2020. High-efficiency air cleaning reduces indoor traffic-related air pollution and alters indoor air chemistry in a near-roadway school. *Environmental Science & Technology* 54(19):11798–808.

Laumbach RJ, Fiedler N, Gardner CR, Laskin DL, Fan ZH, Zhang J, Weschler CJ, Lioy PJ, Devlin RB, Ohman-Strickland P, and 2 others. 2005. Nasal effects of a mixture of volatile organic compounds and their ozone oxidation products. *Journal of Occupational and Environmental Medicine* 47(11):1182–89.

Lee D, Wallis C, Wexler AS, Schelegle ES, Van Winkle LS, Plopper CG, Fanucchi MV, Kumfer B, Kennedy IM, Chan JK. 2010. Small particles disrupt postnatal airway development. *Journal of Applied Physiology* 109(4):1115–24.

Lee LA, Pringle KJ, Reddington CL, Mann GW, Stier P, Spracklen DV, Pierce JR, Carslaw KS. 2013. The magnitude and causes of uncertainty in global model simulations of cloud condensation nuclei. *Atmospheric Chemistry and Physics* 13(17):8879–914.

Lehtimäki M, Heinonen K. 1994. Reliability of electret filters. *Building and Environment* 29(3):353–55.

Lehtimäki M, Taipale A, Saamanen A. 2002. *Investigation of Mechanisms and Operating Environment that Impact the Filtration Efficiency of Charged Air Filtration Media* (Technical Report RP-1189). Atlanta: American Society of Heating, Refrigerating, and Air-Conditioning Engineers.

Less B. 2012. *Indoor Air Quality in 24 California Residences Designed as High Performance Green Homes*. Thesis submitted in partial satisfaction of the requirements for Master of Science in Architecture, University of California, Berkeley.

Li T, Siegel JA. 2020. In situ efficiency of filters in residential central HVAC systems. *Indoor Air* 30(2):315–25.

Li X, Xiao M, Xu X, Zhou J, Yang K, Wang Z, Zhang W, Hopke PK, Zhao W. 2020. Light absorption properties of organic aerosol from wood pyrolysis: Measurement method comparison and radiative implications. *Environmental Science & Technology* 54(12):7156–64.

Licina D, Tian Y, Nazaroff WW. 2017. Emission rates and the personal cloud effect associated with particle release from the perihuman environment. *Indoor Air* 27(4):791–802.

Lim SS, Vos T, Flaxman AD, Danaei G, Shibuya K, Adair-Rohani H, Amann M, Anderson HR, Andrews KG, Aryee M, and 202 others. 2012. A comparative risk assessment of burden of disease and injury attributable to 67 risk factors and risk factor clusters in 21 regions, 1990-2010: A systematic analysis for the Global Burden of Disease study 2010. *The Lancet* 380(9859):2224–60.

Lin K, Marr LC. 2020. Humidity-dependent decay of viruses, but not bacteria, in aerosols and droplets follows disinfection kinetics. *Environmental Science & Technology* 54(2):1024–32.

Lin H, Ma W, Qiu H, Wang X, Trevathan E, Yao Z, Dong GH, Vaughn MG, Qian Z, Tian L. 2017. Using daily excessive concentration hours to explore the short-term mortality effects of ambient $PM_{2.5}$ in Hong Kong. *Environmental Pollution* 229:896–901.

Lin K, Schulte CR, Marr LC. 2020. Survival of MS2 and 6 viruses in droplets as a function of relative humidity, pH, and salt, protein, and surfactant concentrations. *PLOS ONE* 15(12):e0243505.

Lindaas J, Pollack IB, Garofalo LA, Pothier MA, Farmer DK, Kreidenweis SM, Campos TL, Flocke F, Weinheimer AJ, Montzka DD, and 16 others. 2021. Emissions of reactive nitrogen from western U.S. wildfires during summer 2018. *Journal of Geophysical Research: Atmospheres* 126(2):e2020JD032657.

Liu D-L, Nazaroff W. 2001. Modeling pollutant penetration across building envelopes. *Atmospheric Environment* 35:4451–62.

Liu Y, Misztal PK, Xiong J, Tian Y, Arata C, Weber RJ, Nazaroff WW, Goldstein AH. 2019. Characterizing sources and emissions of volatile organic compounds in a northern California residence using space- and time-resolved measurements. *Indoor Air* 29(4):630–44.

Logue JM, McKone TE, Sherman MH, Singer BC. 2011. Hazard assessment of chemical air contaminants measured in residences. *Indoor Air* 21(2):92–109.

Long CM, Suh HH, Kobzik L, Catalano PJ, Ning YY, Koutrakis P. 2001. A pilot investigation of the relative toxicity of indoor and outdoor fine particles: In vitro effects of endotoxin and other particulate properties. *Environmental Health Perspectives* 109(10):1019–26.

Lu KD, Breysse PN, Diette GB, Curtin-Brosnan J, Aloe C, D'Ann LW, Peng RD, McCormack MC, Matsui EC. 2013. Being overweight increases susceptibility to indoor pollutants among urban children with asthma. *Journal of Allergy and Clinical Immunology* 131(4):1017–23.

Lu F, Xu D, Cheng Y, Dong S, Guo C, Jiang X, Zheng X. 2015. Systematic review and meta-analysis of the adverse health effects of ambient $PM_{2.5}$ and PM_{10} pollution in the Chinese population. *Environmental Research* 136:196–204.

Lunden MM, Delp WW, Singer BC. 2015. Capture efficiency of cooking-related fine and ultrafine particles by residential exhaust hoods. *Indoor Air* 25(1):45–58.

Lunderberg DM, Kristensen K, Tian Y, Arata C, Misztal PK, Liu Y, Kreisberg N, Katz EF, DeCarlo PF, Patel S, and 3 others. 2020. Surface emissions modulate indoor SVOC concentrations through volatility-dependent partitioning. *Environmental Science & Technology* 54(11):6751–60.

Madureira J, Alvim-Ferraz MC, Rodrigues S, Gonçalves C, Azevedo MC, Pinto E, Mayan O. 2009. Indoor air quality in schools and health symptoms among Portuguese teachers. *Human and Ecological Risk Assessment* 15(1):159–69.

Majd E, McCormack M, Davis M, Curriero F, Berman J, Connolly F, Leaf P, Rule A, Green T, Clemons-Erby D, and 2 others. 2019. Indoor air quality in inner-city schools and its associations with building characteristics and environmental factors. *Environmental Research* 170:83–91.

Marr LC, Tang JW, Van Mullekom J, Lakdawala SS. 2019. Mechanistic insights into the effect of humidity on airborne influenza virus survival, transmission and incidence. *Journal of the Royal Society Interface* 16(150):20180298.

Martenies SE, Batterman SA. 2018. Effectiveness of using enhanced filters in schools and homes to reduce indoor exposures to $PM_{2.5}$ from outdoor sources and subsequent health benefits for children with asthma. *Environmental Science & Technology* 52(18):10767–776.

Martenies, SE, Milando CW, Batterman SA. 2018. Air pollutant strategies to reduce adverse health impacts and health inequalities: A quantitative assessment for Detroit, Michigan. *Air Quality, Atmosphere & Health* 11(4):409–22.

Mattila JM, Lakey PS, Shiraiwa M, Wang C, Abbatt JP, Arata C, Goldstein AH, Ampollini L, Katz EF, DeCarlo PF, and 8 others. 2020. Multiphase chemistry controls inorganic chlorinated and nitrogenated compounds in indoor air during bleach cleaning. *Environmental Science & Technology* 54(3):1730–39.

McCarthy MC, Ludwig JF, Brown SG, Vaughn DL, Roberts PT. 2013. Filtration effectiveness of HVAC systems at near-roadway schools. *Indoor Air* 23(3):196–207.

McClure CD, Jaffe DA. 2018. US particulate matter air quality improves except in wildfire-prone areas. *Proceedings of the National Academy of Sciences* 115(31):7901–06.

McCormack MC, Breysse PN, Matsui EC, Hansel NN, Williams DA, Curtin-Brosnan J, Eggleston P, Diette GB, and the Center for Childhood Asthma in the Urban Environment. 2009. In-home particle concentrations and childhood asthma morbidity. *Environmental Health Perspectives* 117(2):294–98.

McCormack MC, Belli AJ, Kaji DA, Matsui EC, Brigham EP, Peng RD, Sellers C, D'Ann LW, Diette GB, Breysse PN, Hansel NN. 2015. Obesity as a susceptibility factor to indoor particulate matter health effects in COPD. *European Respiratory Journal* 45(5):1248–57.

McCormack MC, Belli AJ, Waugh D, Matsui EC, Peng RD, Williams DA, Paulin L, Saha A, Aloe CM, Diette GB, and 2 others. 2016. Respiratory effects of indoor heat and the interaction with air pollution in chronic obstructive pulmonary disease. *Annals of the American Thoracic Society* 13(12):2125–31.

Melikov A, Kaczmarczyk J. 2012. Air movement and perceived air quality. *Building and Environment* 47:400–09.

Mendell MJ, Mirer AG, Cheung K, Tong M, Douwes J. 2011. Respiratory and allergic health effects of dampness, mold, and dampness-related agents: A review of the epidemiologic evidence. *Environmental Health Perspectives* 119(6):748–56.

Mendell MJ, Eliseeva EA, Davies MM, Spears M, Lobscheid A, Fisk WJ, Apte MG. 2013. Association of classroom ventilation with reduced illness absence: A prospective study in California elementary schools. *Indoor Air* 23(6):515–28.

Mendoza DL, Benney TM, Boll S. 2021. Long-term analysis of the relationships between indoor and outdoor fine particulate pollution: A case study using research grade sensors. *Science of the Total Environment* 776:145778.

Meng QY, Turpin BJ, Korn L, Weisel CP, Morandi M, Colome S, Zhang J, Stock T, Spektor D, Winer A, and 10 others. 2005. Influence of ambient (outdoor) sources on residential indoor and personal $PM_{2.5}$ concentrations: Analyses of RIOPA data. *Journal of Exposure Science & Environmental Epidemiology* 15(1):17–28.

Michie S, West R, Sheals K, Godinho CA. 2018. Evaluating the effectiveness of behavior change techniques in health-related behavior: A scoping review of methods used. *Translational Behavioral Medicine* 8(2):212–24.

Mikati I, Benson AF, Luben TJ, Sacks JD, Richmond-Bryant J. 2018. Disparities in distribution of particulate matter emission sources by race and poverty status. *American Journal of Public Health* 108(4):480–85.

Mikhailov E, Vlasenko S, Niessner R, Pöschl U. 2004. Interaction of aerosol particles composed of protein and salts with water vapor: Hygroscopic growth and microstructural rearrangement. *Atmospheric Chemistry and Physics* 4(2):323–50.

Morawska L, Thai PK, Liu X, Asumadu-Sakyi A, Ayoko G, Bartonova A, Bedini A, Chai F, Christensen B, Dunbabin M, and 16 others. 2018. Applications of low-cost sensing technologies for air quality monitoring and exposure assessment: How far have they gone? *Environment International* 116:286–99.

Mott JA, Meyer P, Mannino D, Redd SC, Smith EM, Gotway-Crawford C, Chase E. 2002. Wildland forest fire smoke: Health effects and intervention evaluation, Hoopa, California, 1999. *Western Journal of Medicine* 176(3):157–62.

Mukherjee A, Stanton LG, Graham AR, Roberts PT. 2017. Assessing the utility of low-cost particulate matter sensors over a 12-week period in the Cuyama Valley of California. *Sensors* 17(8):1805.

NASEM [National Academies of Sciences, Engineering, and Medicine]. 2016. *Health Risks of Indoor Exposure to Particulate Matter: Workshop Summary*. Washington: National Academies Press.

NASEM. 2017. *Microbiomes of the Built Environment: A Research Agenda for Indoor Microbiology, Human Health, and Buildings*. Washington: National Academies Press.

Newton A, Serdar B, Adams K, Dickinson LM, Koehler K. 2021. Lung deposition versus inhalable sampling to estimate body burden of welding fume exposure: a pilot sampler study in stainless steel welders. *Journal of Aerosol Science* 153:105721.

Nguyen JL, Dockery DW. 2016. Daily indoor-to-outdoor temperature and humidity relationships: A sample across seasons and diverse climatic regions. *International Journal of Biometeorology* 60(2):221–29.

Nguyen JL, Schwartz J, Dockery DW. 2014. The relationship between indoor and outdoor temperature, apparent temperature, relative humidity, and absolute humidity. *Indoor Air* 24(1):103–12.

Novoselac A, Siegel J. 2009. Impact of placement of portable air cleaning devices in multizone residential environments. *Building and Environment* 44:2348–56.

O'Brien RE, Li Y, Kiland KJ, Katz EF, Or VW, Legaard E, Walhout EQ, Thrasher C, Grassian VH, DeCarlo PF, and 2 others. 2021. Emerging investigator series: Chemical and physical properties of organic mixtures on indoor surfaces during HOMEChem. *Environmental Science: Processes & Impacts* 23(4):559–68.

Oeder S, Jörres RA, Weichenmeier I, Pusch G, Schober W, Pfab F, Behrendt H, Schierl R, Kronseder A, Nowak D, and 8 others. 2012. Airborne indoor particles from schools are more toxic than outdoor particles. *American Journal of Respiratory Cell Molecular Biology* 47(5):575–82.

Or VW, Wade M, Patel S, Alves MR, Kim D, Schwab S, Przelomski H, O'Brien R, Rim D, Corsi RL, Vance ME, and 2 others. 2020. Glass surface evolution following gas adsorption and particle deposition from indoor cooking events as probed by microspectroscopic analysis. *Environmental Science: Processes & Impacts* 22(8):1698–709.

Palm BB, Peng Q, Fredrickson CD, Lee BH, Garofalo LA, Pothier MA, Kreidenweis SM, Farmer DK, Pokhrel RP, Shen Y, and 10 others. 2020. Quantification of organic aerosol and brown carbon evolution in fresh wildfire plumes. *Proceedings of the National Academy of Sciences* 117(47): 29469–77.

Pan J, Tang J, Caniza M, Heraud JM, Koay E, Lee HK, Lee CK, Li Y, Nava Ruiz A, Santillan-Salas CF, Marr LC. 2021. Correlating indoor and outdoor temperature and humidity in a sample of buildings in tropical climates. *Indoor Air* 31(6):2281–95.

Park JH, Lee TJ, Park MJ, Oh HN, Jo YM. 2020. Effects of air cleaners and school characteristics on classroom concentrations of particulate matter in 34 elementary schools in Korea. *Building and Environment* 167:106437.

Parker JD, Kravets N, Vaidyanathan A. 2018. Particulate matter air pollution exposure and heart disease mortality risks by race and ethnicity in the United States. *Circulation* 137(16):1688–97.

Patel S, Sankhyan S, Boedicker EK, DeCarlo PF, Farmer DK, Goldstein AH, Katz EF, Nazaroff WW, Tian Y, Vanhanen J, Vance ME. 2020. Indoor particulate matter during HOMEChem: Concentrations, size distributions, and exposures. *Environmental Science & Technology* 54(12):7107–16.

Peng C, Ni P, Xi G, Tian W, Fan L, Zhou D, Zhang Q, Tang Y. 2020. Evaluation of particle penetration factors based on indoor $PM_{2.5}$ removal by an air cleaner. *Environmental Science and Pollution Research* 27(8):8395–405.

Phillips TJ. 2019. Ducted range hoods: Recommendations for new and existing homes. *Reducing Outdoor Contaminants in Indoor Spaces*. ROCIS Issue Brief, https://rocis.org/range-hood-document (accessed August 1, 2021).

Phipatanakul W, Bailey A, Hoffman EB, Sheehan WJ, Lane JP, Baxi S, Rao D, Permaul P, Gaffin JM, Rogers CA, and 2 others. 2011. The School Inner-City Asthma Study: Design, methods, and lessons learned. *Journal of Asthma* 48(10):1007–14.

Pirela S, Molina R, Watson C, Cohen JM, Bello D, Demokritou P, Brain J. 2013. Effects of copy center particles on the lungs: A toxicological characterization using a BALB/c mouse model. *Inhalation Toxicology* 25(9):498–508.

Polidori A, Fine PM, White V, Kwon PS. 2013. Pilot study of high-performance air filtration for classroom applications. *Indoor Air* 23(3):185–95.

Prussin AJ, Marr LC. 2015. Sources of airborne microorganisms in the built environment. *Microbiome* 3(1):78.

Qian J, Hospodsky D, Yamamoto N, Nazaroff WW, Peccia J. 2012. Size-resolved emission rates of airborne bacteria and fungi in an occupied classroom. *Indoor Air* 22(4):339–51.

Raju S, Brigham EP, Paulin LM, Putcha N, Balasubramanian A, Hansel NN, McCormack MC. 2019a. The burden of rural chronic obstructive pulmonary disease: Analyses from the National Health and Nutrition Examination Survey. *American Journal of Respiratory and Critical Care Medicine* 201(4):488–91.

Raju S, Keet CA, Paulin LM, Matsui EC, Peng RD, Hansel NN, McCormack MC. 2019b. Rural residence and poverty are independent risk factors for chronic obstructive pulmonary disease in the United States. *American Journal of Respiratory and Critical Care Medicine* 199(8):961–69.

Reisen F, Powell JC, Dennekamp M, Johnston FH, Wheeler AJ. 2019. Is remaining indoors an effective way of reducing exposure to fine particulate matter during biomass burning events? *Journal of the Air & Waste Management Association* 69(5):611–22.

Rich DQ, Freudenberger RS, Ohman-Strickland P, Cho Y, Kipen HM. 2008. Right heart pressure increases after acute increases in ambient particulate concentration. *Environmental Health Perspectives* 116(9):1167–71.

Rich DQ, Kipen HM, Huang W, Wang G, Wang Y, Zhu P, Ohman-Strickland P, Hu M, Philipp C, Diehl SR, and 6 others. 2012. Association between changes in air pollution levels during the Beijing Olympics and biomarkers of inflammation and thrombosis in healthy young adults. *Journal of the American Medical Association* 307(19):2068–78.

Rim D, Wallace L, Persily A. 2010. Infiltration of outdoor ultrafine particles into a test house. *Environmental Science & Technology* 44(15):5908–13.

Rogalsky DK, Mendola P, Metts TA, Martin WJ. 2014. Estimating the number of low-income Americans exposed to household air pollution from burning solid fuels. *Environmental Health Perspectives* 122(8):806–10.

Rongerude J, Haddad M. 2016. Cores and peripheries: Spatial analysis of housing choice voucher distribution in the San Francisco Bay Area region, 2000–2010. *Housing Policy Debate* 26(3):417–36.

Sampson A, Holgate S. 1998. Leukotriene modifiers in the treatment of asthma. *British Medical Journal* 316(7140):1257–58.

Sarnat JA, Long CM, Koutrakis P, Coull BA, Schwartz J, Suh HH. 2002. Using sulfur as a tracer of outdoor fine particulate matter. *Environmental Science & Technology* 36(24):5305–14.

Sarwar G, Corsi R, Allen D, Weschler C. 2003. The significance of secondary organic aerosol formation and growth in buildings: Experimental and computational evidence. *Atmospheric Environment* 37:1365–81.

Scheepers PTJ, de Hartog JJ, Reijnaerts J, Beckmann G, Anzion R, Poels K, Godderis L. 2015. Influence of combined dust reducing carpet and compact air filtration unit on the indoor air quality of a classroom. *Environmental Science: Processes & Impacts* 17(2):316–25.

Schiavon S, Yang B, Donner Y, Chang VC, Nazaroff WW. 2017. Thermal comfort, perceived air quality, and cognitive performance when personally controlled air movement is used by tropically acclimatized persons. *Indoor Air* 27(3):690–702.

Schraufnagel DE. 2020. The health effects of ultrafine particles. *Experimental & Molecular Medicine* 52(3):311–17.

Sexton K, Selevan SG, Wagener DK, Lybarger JA. 1992. Estimating human exposures to environmental pollutants: Availability and utility of existing databases. *Archives of Environmental Health* 47(6):398–407.

Shang J, Khuzestani RB, Tian J, Schauer JJ, Hua J, Zhang Y, Cai T, Fang D, An J, Zhang Y. 2019. Chemical characterization and source apportionment of $PM_{2.5}$ personal exposure of two cohorts living in urban and suburban Beijing. *Environmental Pollution* 246:225–36.

Singer BC, Delp WW, Price PN, Apte MG. 2012. Performance of installed cooking exhaust devices. *Indoor Air* 22(3):224–34.

Singer BC, Delp WW, Black DR, Walker IS. 2017a. Measured performance of filtration and ventilation systems for fine and ultrafine particles and ozone in an unoccupied modern California house. *Indoor Air* 27(4):780–90.

Singer BC, Pass RZ, Delp WW, Lorenzetti DM, Maddalena RL. 2017b. Pollutant concentrations and emission rates from natural gas cooking burners without and with range hood exhaust in nine California homes. *Building and Environment* 122:215–29.

Smith KR, Bruce N, Balakrishnan K, Adair-Rohani H, Balmes J, Chafe Z, Dherani M, Hosgood HD, Mehta S, Pope D, Rehfuess E. 2014. Millions dead: How do we know and what does it mean? Methods used in the comparative risk assessment of household air pollution. *Annual Review of Public Health* 35(1):185–206.

Stauffer DA, Autenrieth DA, Hart JF, Capoccia S. 2020. Control of wildfire-sourced $PM_{2.5}$ in an office setting using a commercially available portable air cleaner. *Journal of Occupational and Environmental Hygiene* 17(4):109–20.

Stephens B. 2015. Building design and operational choices that impact indoor exposures to outdoor particulate matter inside residences. *Science and Technology for the Built Environment* 21(1):3–13.

Stephens B, Siegel JA, Novoselac A. 2011. Operational characteristics of residential and light-commercial air-conditioning systems in a hot and humid climate zone. *Building and Environment* 46(10):1972–83.

Sun L, Wallace LA. 2021. Residential cooking and use of kitchen ventilation: The impact on exposure. *Journal of the Air & Waste Management Association* 71(7):830–43.

Sun YL, Zhang Q, Schwab JJ, Chen WN, Bae MS, Hung HM, Lin YC, Ng NL, Jayne J, Massoli P, and 2 others 2012. Characterization of near-highway submicron aerosols in New York City with a high-resolution aerosol mass spectrometer. *Atmospheric Chemistry and Physics* 12(4):2215–27.

Sun L, Wallace LA, Dobbin NA, You H, Kulka R, Shin T, St-Jean M, Aubin D, Singer BC. 2018. Effect of venting range hood flow rate on size-resolved ultrafine particle concentrations from gas stove cooking. *Aerosol Science and Technology* 52(12):1370–81.

Sunil VR, Laumbach RJ, Patel KJ, Turpin BJ, Lim HJ, Kipen HM, Laskin JD, Laskin DL. 2007. Pulmonary effects of inhaled limonene ozone reaction products in elderly rats. *Toxicology and Applied Pharmacology* 222(2):211–20.

Sunyer J, Esnaola M, Alvarez-Pedrerol M, Forns J, Rivas I, López-Vicente M, Suades-González E, Foraster M, Garcia-Esteban R, Basagaña X, and 7 others. 2015. Association between traffic-related air pollution in schools and cognitive development in primary school children: A prospective cohort study. *PLOS Medicine* 12(3):e1001792.

Tang CH, Garshick E, Grady S, Coull B, Schwartz J, Koutrakis P. 2018. Development of a modeling approach to estimate indoor-to-outdoor sulfur ratios and predict indoor $PM_{2.5}$ and black carbon concentrations for eastern Massachusetts households. *Journal of Exposure Science & Environmental Epidemiology* 28(2):125–30.

Thatcher TL, Lunden MM, Revzan KL, Sextro RG, Brown NJ. 2003. A concentration rebound method for measuring particle penetration and deposition in the indoor environment. *Aerosol Science and Technology* 37(11):847–64.

Thurston GD, Kipen H, Annesi-Maesano I, Balmes J, Brook RD, Cromar K, De Matteis S, Forastiere F, Forsberg B, Frampton MW, and 14 others. 2017. A joint ERS/ATS policy statement: What constitutes an adverse health effect of air pollution? An analytical framework. *European Respiratory Journal* 49(1):1600419.

Torkmahalleh A, Gorjinezhad M, Unluevcek A, Sumru H, Hopke PK. 2017. Review of factors impacting emission/concentration of cooking generated particulate matter. *Science of the Total Environment* 586:1046–56.

Touchie MF, Siegel JA. 2018. Residential HVAC runtime from smart thermostats: Characterization, comparison, and impacts. *Indoor Air* 28(6):905–15.

US EPA [US Environmental Protection Agency]. 2019. *Integrated Science Assessment (ISA) for Particulate Matter* (Final Report, Dec 2019). EPA/600/R-19/188. Washington, DC.

US EPA. 2021. *Air Cleaners and Air Filters in the Home*. Washington, DC. https://www.epa.gov/indoor-air-quality-iaq/air-cleaners-and-air-filters-home (accessed October 1, 2021).

van der Zee SC, Strak M, Dijkema MB, Brunekreef B, Janssen NA. 2017. The impact of particle filtration on indoor air quality in a classroom near a highway. *Indoor Air* 27(2):291–302.

VerShaw J, Siegel JA, Chojnowski DB, Nigro PJ. 2009. Implications of filter bypass. *ASHRAE Transactions* 155(1):191–98.

Vette AF, Rea AW, Lawless PA, Rodes CE, Evans G, Highsmith VR, Sheldon L. 2001. Characterization of indoor-outdoor aerosol concentration relationships during the Fresno PM exposure studies. *Aerosol Science and Technology* 34(1):118–26.

Wallace L. 1996. Indoor particles: A review. *Journal of the Air & Waste Management Association* 46(2):98–126.

Wallace L, Williams R. 2005. Use of personal-indoor-outdoor sulfur concentrations to estimate the infiltration factor and outdoor exposure factor for individual homes and persons. *Environmental Science & Technology* 39(6):1707–14.

Wang C, Waring MS. 2014. Secondary organic aerosol formation initiated from reactions between ozone and surface-sorbed squalene. *Atmospheric Environment* 84:222–29.

Wang Y, Li J, Jing H, Zhang Q, Jiang J, Biswas P. 2015. Laboratory evaluation and calibration of three low-cost particle sensors for particulate matter measurement. *Aerosol Science and Technology* 49(11):1063–77.

Wang C, Collins DB, Abbatt JPD. 2019. Indoor illumination of terpenes and bleach emissions leads to particle formation and growth. *Environmental Science & Technology* 53(20):11792–800.

Ward M, Siegel J. 2005. Modeling filter bypass: Impact on filter efficiency. *ASHRAE Transactions* 111(1):1091–100.

Wargocki P, Wyon DP. 2007. The effects of moderately raised classroom temperatures and classroom ventilation rate on the performance of schoolwork by children (RP–1257). *HVAC&R Research* 13(2):193–220.

Waring MS. 2014. Secondary organic aerosol in residences: Predicting its fraction of fine particle mass and determinants of formation strength. *Indoor Air* 24(4):376–89.

Wellenius GA, Schwartz J, Mittleman MA. 2006. Particulate air pollution and hospital admissions for congestive heart failure in seven United States cities. *American Journal of Cardiology* 97(3):404–08.

Weschler C. 2009. Changes in indoor pollutants since the 1950s. *Atmospheric Environment* 43:153–69.

Weschler CJ, Shields HC. 1999. Indoor ozone/terpene reactions as a source of indoor particles. *Atmospheric Environment* 33(15):2301–12.

Yao W, Gallagher DL, Marr LC, Dietrich AM. 2019. Emission of iron and aluminum oxide particles from ultrasonic humidifiers and potential for inhalation. *Water Research* 164:114899.

Yao W, Dal Porto R, Gallagher DL, Dietrich AM. 2020. Human exposure to particles at the air-water interface: Influence of water quality on indoor air quality from use of ultrasonic humidifiers. *Environment International* 143:105902.

Yeh H-C, Cuddihy RG, Phalen RF, Chang IY. 1996. Comparisons of calculated respiratory tract deposition of particles based on the proposed NCRP model and the new ICRP66 model. *Aerosol Science and Technology* 25(2):134–40.

Youssefi S, Waring MS. 2012. Predicting secondary organic aerosol formation from terpenoid ozonolysis with varying yields in indoor environments. *Indoor Air* 22(5):415–26.

Yu X, Stuart AL, Liu Y, Ivey CE, Russell AG, Kan H, Henneman LR, Sarnat SE, Hasan S, Sadmani A, and 2 others. 2019. On the accuracy and potential of Google Maps location history data to characterize individual mobility for air pollution health studies. *Environmental Pollution* 252:924–30.

Yu X, Ivey C, Huang Z, Gurram S, Sivaraman V, Shen H, Eluru N, Hasan S, Henneman L, Shi G, and 2 others. 2020. Quantifying the impact of daily mobility on errors in air pollution exposure estimation using mobile phone location data. *Environment International* 141:105772.

Yuan Y, Luo Z, Liu J, Wang Y, Lin Y. 2018. Health and economic benefits of building ventilation interventions for reducing indoor $PM_{2.5}$ exposure from both indoor and outdoor origins in urban Beijing, China. *Science of the Total Environment* 626:546–54.

Zeng Y, Yu H, Zhao H, Stephens B, Verma V. 2021. Influence of environmental conditions on the dithiothreitol (DTT)-based oxidative potential of size-resolved indoor particulate matter of ambient origin. *Atmospheric Environment* 255:118429.

Zhang Q, Jimenez JL, Canagaratna MR, Ulbrich IM, Ng NL, Worsnop DR, Sun Y. 2011. Understanding atmospheric organic aerosols via factor analysis of aerosol mass spectrometry: A review. *Analytical and Bioanalytical Chemistry* 401(10):3045–67.

Zhang J, Zhu T, Kipen H, Wang G, Huang W, Rich D, Zhu P, Wang Y, Lu SE, Ohman-Strickland P, and 5 others. 2013. Cardiorespiratory biomarker responses in healthy young adults to drastic air quality changes surrounding the 2008 Beijing Olympics. *Research Reports: Health Effects Institute* (174):5.

Zhang Y, Li T, Siegel JA. 2020. Investigating the impact of filters on long-term particle concentration measurements in residences (RP–1649). *Science and Technology for the Built Environment* 26(8):1037–47.

Zhao H, Stephens B. 2017. Using portable particle sizing instrumentation to rapidly measure the penetration of fine and ultrafine particles in unoccupied residences. *Indoor Air* 27(1):218–29.

Zhou J, Chen A, Cao Q, Yang B, Chang VWC, Nazaroff WW. 2015. Particle exposure during the 2013 haze in Singapore: Importance of the built environment. *Building and Environment* 93:14–23.

Zhu Y, Hinds WC, Kim S, Shen S, Sioutas C. 2002. Study of ultrafine particles near a major highway with heavy-duty diesel traffic. *Atmospheric Environment* 36(27):4323–35.

Zou Y, Young M, Wickey M, May A, Clark JD. 2020. Response of eight low-cost particle sensors and consumer devices to typical indoor emission events in a real home (ASHRAE 1756-RP). *Science and Technology for the Built Environment* 26(2):237–49.

Appendix A

Workshop Agenda

WORKSHOP ON INDOOR EXPOSURE TO FINE PARTICULATE MATTER AND PRACTICAL MITIGATION APPROACHES

APRIL 14, 21, AND 28, 2021

APRIL 14 WEBINAR – SOURCES OF INDOOR FINE PARTICULATE MATTER

11:00 am **Welcome; workshop and session goals**
Richard Corsi, PhD, PE – Planning Committee Chair

11:10 am **Sponsor remarks**
Jonathan Edwards
Director, Office of Radiation and Indoor Air,
US Environmental Protection Agency

SESSION I: OUTDOOR SOURCES OF INDOOR PARTICULATE MATTER

11:15 am **Introduction of session speakers**
Kimberly Prather, PhD – Session Moderator and Planning Committee Member

11:20 am	**Indoor Particulate Matter of Outdoor Origin and the Disparities in Sources and Exposures Across Communities** *Cesunica Ivey, PhD* Assistant Professor, Chemical/Environmental Engineering, University of California, Riverside
11:40 am	**Outdoor-to-Indoor Transport Mechanisms and Particle Penetration for Fine Particulate Matter** *Brent Stephens, PhD* Professor and Department Chair, Department of Civil, Architectural, and Environmental Engineering, Illinois Institute of Technology
12:00 pm	**Outdoor Particulate Matter Sources and the Chemical Transformations that Take Place When They Interact with the Indoor Environment** *Delphine Farmer, PhD* Associate Professor, Department of Chemistry, Colorado State University
12:20 pm	**Roundtable Discussion** Session Speakers and Planning Committee Members
12:45 pm	**Break**

SESSION II: INDOOR SOURCES OF INDOOR PARTICULATE MATTER

12:55 pm	**Introduction of session speakers** *Kimberly Prather, PhD – Session Moderator and Planning Committee Member*
1:00 pm	**Fine Particulate Matter Emissions From Cooking** *Marina Vance, PhD* Assistant Professor and McLagan Family Faculty Fellow, Department of Mechanical Engineering, University of Colorado Boulder

1:20 pm	**Secondary Aerosol Formation of Fine Particulate Matter in the Indoor Environment** *Michael Waring, PhD* Department Head and Professor, Department of Civil, Architectural and Environmental Engineering, Drexel University College of Engineering
1:40 pm	**The Effect of Humidity on the Chemistry and Biology of Indoor Air** *Linsey Marr, PhD* Charles P. Lunsford Professor of Civil and Environmental Engineering, Virginia Tech
2:00 pm	**The Influence of Sources of Indoor Fine Particulate Matter on the Characterization of Exposure and Evaluation of Health Effects** *Andrea Ferro, PhD* Professor / ISE Associate Director for Research, Department of Civil & Environmental Engineering, Clarkson University
2:20 pm	**Roundtable Discussion** Session Speakers and Planning Committee Members
2:50 pm	**Session wrap-up and preview of upcoming webinars** *Kimberly Prather, PhD – Session Moderator and Planning Committee Member*
3:00 pm	**Session adjourns**

APRIL 21 WEBINAR – INDOOR EXPOSURE TO FINE PARTICULATE MATTER: HEALTH, METRICS, AND ASSESSMENT

11:00 am	**Welcome; workshop and session goals** *Richard Corsi, PhD, PE – Planning Committee Chair*
11:10 am	**Brief summary of the previous workshop session**

SESSION I: HEALTH EFFECTS OF EXPOSURE TO INDOOR PARTICULATE MATTER

11:15 am   **Introduction of session speakers**
Elizabeth Matsui, MD, MHS – Session Moderator and Planning Committee Member

11:20 am   **The Overall (Mostly Cardiovascular) Health Burden of Indoor $PM_{2.5}$ Exposure**
Howard Kipen, MD, MPH
Professor, Department of Occupational and Environmental Health, Rutgers University School of Public Health

11:45 am   **Pulmonary Disease Associated with Fine Particulate Matter Exposure in Indoor Environments and Disparities in Economically Challenged Communities**
Meredith McCormack, MD, MHS
Medical Director, Pulmonary Function Laboratory and Associate Professor of Medicine, Johns Hopkins University School of Medicine

12:05 pm   **Wildfire Smoke and Other Ambient Air Pollution Comes Indoors: Health Effects and the Building Characteristics That Mitigate Them**
Stephanie Holm, MD, MPH
Co-director, Western States Pediatric Environmental Health Specialty Unit, University of California, San Francisco

12:25 pm   **Moderated roundtable discussion**
Session Speakers and Planning Committee Members

Elizabeth Matsui, MD, MHS, and Linda A. McCauley, PhD, RN, FAAN, FAAOHN – Comoderators

12:50 pm   **Break**

SESSION II: INDOOR EXPOSURE TO PARTICULATE MATTER: METRICS AND ASSESSMENT

1:00 pm   **Introduction of session speakers**
Elizabeth Matsui, MD, MHS – Session Moderator and Planning Committee Member

APPENDIX A 143

1:05 pm	**Transcending Complexity: Indoor Fine Particulate Matter Measurement, Exposure, and Control** *William W. Nazaroff, PhD* Daniel Tellep Distinguished Professor Emeritus, Department of Civil and Environmental Engineering, University of California, Berkeley
1:30 pm	**The Challenge of Moving from the Measurement of Fine Indoor Particulate Matter to Evaluating Occupant Exposure** *Kirsten Koehler, PhD* Associate Professor, Johns Hopkins Bloomberg School of Public Health
1:50 pm	**The Utility, Use, and Misuse of Low-Cost Consumer Indoor Particulate Matter Sensors** *Dusan Licina, PhD* Assistant Professor, Indoor Environmental Quality, School for Architecture, Civil, and Environmental Engineering, Swiss Federal Institute of Technology, Lausanne
2:10 pm	**Moderated roundtable discussion** Session speakers and Planning Committee members *Elizabeth Matsui, MD, MHS, and Linda A. McCauley, PhD, RN, FAAN, FAAOHN – Comoderators*
2:50 pm	**Session wrap-up and preview of upcoming webinar** *Seema Bhangar, PhD – Session Moderator and Planning Committee Member*
3:00 pm	**Session adjourns**

APRIL 28 WEBINAR – MITIGATION OF INDOOR EXPOSURE TO FINE PARTICULATE MATTER

11:00 am	**Welcome; workshop and session goals; summary of the previous workshop sessions** *Richard Corsi, PhD, PE – Planning Committee Chair*

SESSION I: INDOOR PARTICULATE MATTER EXPOSURE CONTROL AND MITIGATION

11:10 am **Introduction of session speakers**
Wanyu (Rengie) Chan, PhD – Session Moderator and Planning Committee Member

11:15 am **Fine Particulate Matter Filtration and Air Cleaning in Residential Environments**
Jeffrey Siegel, PhD
Professor of Civil Engineering, University of Toronto

11:35 am **Fine Particulate Matter Exposure Control in Schools**
Elliott Gall, PhD
Assistant Professor, Department of Mechanical and Materials Engineering, Maseeh College of Engineering and Computer Science, Portland State University

11:55 am **Mitigation of Fine Particulate Matter Exposures Associated with Cooking**
Brett Singer, PhD
Staff Scientist and Principal Investigator, Energy Technologies Area, Lawrence Berkeley National Laboratory

12:15 pm **Moderated roundtable discussion**
Session Speakers and Planning Committee Members
Wanyu (Rengie) Chan, PhD, and Seema Bhangar, PhD – Comoderators

12:50 pm **Break**

SESSION II: OCCUPANT RESPONSES TO INDOOR PARTICULATE MATTER

1:00 pm **Introduction of session speakers**
Wanyu (Rengie) Chan, PhD – Session Moderator and Planning Committee Member

APPENDIX A   *145*

1:05 pm	**Portable Indoor Air Cleaners and Human Behavior** *Stuart Batterman, PhD* Professor, Environmental Health Sciences and Global Public Health, University of Michigan School of Public Health
1:25 pm	**How Building Occupants Interpret and Respond to Indoor Air Quality Sensor Data** *Lindsay Graham, PhD* Research Specialist, Center for the Built Environment, University of California, Berkeley
1:45 pm	**Public Health Responses to Reduce Community Exposure to Indoor Fine Particulate Matter** *Sarah Coefield, MS, MA* Air Quality Specialist, Missoula City-County Health Department
2:05 pm	**Moderated roundtable discussion** Session Speakers and Planning Committee Members *Wanyu (Rengie) Chan, PhD, and Seema Bhangar, PhD – Comoderators*
2:40 pm	**Workshop summary and closing reflections** *Richard Corsi, PhD, PE – Planning Committee Chair*
2:55 pm	**Workshop concludes**

Appendix B

Biographic Sketches of Planning Committee Members and Workshop Speakers

Planning Committee Members

Richard L. Corsi, PhD, PE (Chair), is the dean of engineering at the University of California, Davis. Previously, he was the H. Chik M. Erzurulu Dean of the Maseeh College of Engineering and Computer Science at Portland State University and, before that, a faculty member, department chair, and endowed research chair at the University of Texas at Austin in the Department of Civil, Architectural and Environmental Engineering. Dr. Corsi is an internationally recognized expert in the field of indoor air quality, with a specific interest in physical and chemical interactions between pollutants and indoor materials. He and his team have published nearly 270 peer-reviewed papers stemming from 70 funded research projects and supervision of over 120 students in research. He was inducted into the International Society of Indoor Air Quality and Climates' Academy of Fellows in 2008, and currently serves as academy president. Dr. Corsi was a member of the planning committee responsible for the 2016 National Academies report *Health Risks of Indoor Exposure to Particulate Matter: Workshop Summary*. He received his BS degree in environmental resources engineering from Humboldt State University, where he was honored as a distinguished alumnus in 2006, and his MS and PhD degrees in civil engineering from the University of California, Davis, where he received a Distinguished Engineering Alumni Medal in 2016.

Seema Bhangar, PhD, is the senior indoor air quality manager at WeWork, a commercial real estate company that provides flexible workspaces. Her mission is to drive the use of new science and technology for designing, building, and operating buildings that are better for health and productivity, and more resilient and energy efficient. She specializes in indoor air quality research projects with a focus on applying human-centric approaches to environmental sensing in buildings and transportation systems. Dr. Bhangar previously served as technical lead and product manager for the design and development of next-generation indoor sensing devices for Aclima, Inc. She is a regular peer reviewer for journals including *Indoor Air*, *Building and Environment*, and *Environmental Science & Technology*. She earned a BAS from Stanford University and an MS and PhD in environmental engineering from the University of California, Berkeley.

Wanyu (Rengie) Chan, PhD, is a research scientist and deputy leader of the Indoor Environment Group in Energy Analysis and Environmental Impact Division at Lawrence Berkeley National Laboratory. Her work focuses on characterizing indoor air quality and implications for human exposures in residential and commercial buildings. Dr. Chan led a recent field study to evaluate the role of mechanical ventilation on indoor air quality in new California homes, and she is part of an ongoing project funded by the Department of Energy's Building America Program to study indoor air quality in new homes across different US regions. She has also modeled the health benefits from filtration of ambient $PM_{2.5}$ and during wildfire smoke. She joined Berkeley Lab as a graduate student and worked on the evaluation of shelter-in-place effectiveness. Dr. Chan earned her BS in chemical engineering from Carnegie Mellon University and her MS and PhD in civil and environmental engineering from the University of California, Berkeley.

Elizabeth Matsui, MD, MHS, is a professor of population health and pediatrics at the Dell Medical School at the University of Texas at Austin, where she is also director of clinical and translational research. She is a leading international expert on environmental allergies and asthma. Her research focuses on examining the impact of allergen exposure on allergic disease. She serves on the editorial board of the *Journal of Allergy and Clinical Immunology* and is a member of the American College of Allergy, Asthma, and Immunology and the American Academy of Allergy, Asthma, and Immunology. Dr. Matsui served on the National Academies' Standing Committee on Medical and Epidemiological Aspects of Air Pollution on US Government Employees and Their Families. She received her undergraduate degree in molecular biology and her MD from Vanderbilt University, and completed a master of health science in epidemiology at the Johns Hopkins Bloomberg School of Public Health.

Linda A. McCauley, PhD, RN, FAAN, FAAOHN (NAM), is a professor and dean of Emory University's Nell Hodgson Woodruff School of Nursing. She has special knowledge in the design of epidemiological investigations of environmental hazards and is nationally recognized for her expertise in occupational and environmental health nursing. Her work aims to identify culturally appropriate interventions to decrease the impact of environmental and occupational health hazards in vulnerable populations, including workers and young children. Dr. McCauley was previously the associate dean for research and Nightingale Professor in Nursing at the University of Pennsylvania School of Nursing. She has served on numerous National Academies committees. She received a bachelor of nursing degree from the University of North Carolina, a master's in nursing from Emory, and a PhD in environmental health and epidemiology from the University of Cincinnati. Dr. McCauley was elected a Member of the National Academy of Medicine in 2008.

Kimberly A. Prather, PhD (NAE, NAS), holds a joint appointment as professor of chemistry and biochemistry at Scripps Institution of Oceanography and the University of California, San Diego. Her research involves the development and application in field and lab studies of real-time measurements of size-resolved chemistry of aerosols. She is involved in aerosol source apportionment studies and her group is working to better understand the impact of specific aerosol sources on health and climate. Dr. Prather was a member of the Fine Particle Monitoring Subcommittee of EPA's Clean Air Scientific Advisory Committee. She is on a number of editorial boards for journals including *Aerosol Science and Technology*, and is a member of professional societies including the American Association for Aerosol Research, American Chemical Society, and American Geophysical Union. Dr. Prather received her BS and PhD in chemistry from the University of California, Davis. She was elected a Member of the National Academy of Engineering in 2019 and the National Academy of Sciences in 2020.

David Y.H. Pui, PhD (NAE), is Regents Professor and the L.M. Fingerson/ TSI Inc. Chair in Mechanical Engineering at the University of Minnesota, Minneapolis, and director of both the university's Particle Technology Laboratory and Center for Filtration Research. He has a broad range of research experience in aerosol science and technology, including particle instrumentation development and filtration solutions for air pollution control; and development of instrumentation for generating, sampling, and measuring airborne particles. He has authored more than 320 peer-reviewed journal papers and has over 30 patents, and has developed or codeveloped several widely used commercial aerosol instruments. Dr. Pui previously served as president of the American Association for Aerosol

Research and of the International Aerosol Research Assembly, consisting of 17 international aerosol associations. He earned bachelor's, master's and PhD degrees in mechanical engineering from the University of Minnesota. Dr. Pui was elected a Member of the National Academy of Engineering in 2016.

Jeffrey Siegel, PhD,[1] is a professor of civil engineering at the University of Toronto and a member of the university's Building Engineering Research Group. He holds joint appointments at the Dalla Lana School of Public Health and the Department of Physical & Environmental Sciences. Dr. Siegel is a fellow of the American Society of Heating, Refrigerating and Air-Conditioning Engineers (ASHRAE) and a member of the Academy of Fellows of the International Society of Indoor Air Quality and Climate (ISIAQ). His research interests include healthy and sustainable buildings, ventilation and indoor air quality in residential and commercial buildings, control of indoor particulate matter, the indoor microbiome, and moisture interactions with indoor chemistry and biology. Dr. Siegel earned a BSc from Swarthmore College and an MS and PhD in mechanical engineering from the University of California, Berkeley.

Marina Vance, PhD,[1] is an assistant professor and McLagan Family Faculty Fellow in the Department of Mechanical Engineering at the University of Colorado Boulder, and holds a courtesy appointment in the university's Environmental Engineering program. Her research is focused on air quality, particularly on measuring emissions and understanding the dynamics of aerosols in the context of ambient and indoor air quality. She is the principal investigator of the HOMEChem (House Observations of Microbial and Environmental Chemistry) research initiative, which incorporated measurements from over 20 research groups at 13 universities to identify the most important aspects of the chemistry that controls the indoor environment. Dr. Vance earned a BS in sanitation and environmental engineering, an MS in environmental engineering from the Universidade Federal de Santa Catarina (Brazil), and a PhD in civil and environmental engineering from Virginia Tech.

David A. Butler, PhD (National Academies staff officer), is the J. Herbert Hollomon Scholar of the National Academy of Engineering. He previously held an appointment as scholar in the National Academies' Health and Medicine Division. Before joining the National Academies, he was an analyst for the US Congress Office of Technology Assessment, a research associate in the Department of Environmental Health of the Harvard T.H. Chan School of Public Health, a researcher at Harvard's John F. Kennedy

---

[1] Drs. Siegel and Vance also served as workshop speakers.

School of Government, and a product safety engineer for Xerox Corporation. He has directed several National Academies studies on environmental health and risk assessment topics, including those that produced *Clearing the Air: Asthma and Indoor Air Exposures*; *Damp Indoor Spaces and Health*; *Climate Change, the Indoor Environment, and Health*; and *Health Risks of Indoor Exposure to Particulate Matter: Workshop Summary*. Dr. Butler earned his BS and MS in electrical engineering, with a specialization in biomedical engineering, from the University of Rochester and his PhD in public policy analysis from Carnegie Mellon University. He is a recipient of the National Academies' Cecil Medal for Research.

Workshop Speakers

Stuart Batterman, PhD, is a professor in environmental health sciences as well as global public health at the University of Michigan School of Public Health. His research and teaching interests address environmental impact assessment, human exposure and health risk assessment, and environmental management involving both theoretical work and applied laboratory and field studies. He is particularly interested in improving exposure measures that can be used in risk assessments and epidemiological studies; measuring toxic compounds including volatile organic compounds (VOCs) found as pollutants in drinking water, ambient and indoor air; and statistical and modeling methods that can be used to interpret and extend available measurements. His research is applied to contemporary problems including ambient and indoor air quality, environmental epidemiology, policy analysis, environmental engineering, environmental justice, and life cycle analysis. Dr. Batterman earned a BS in environmental sciences from Rutgers University, and an MS and PhD in water resources and environmental engineering from the Massachusetts Institute of Technology.

Sarah Coefield, MS, MA, is an Air Quality Specialist with the Missoula City-County Health Department. As a public health practitioner she works to prepare the public for air pollution events through studies, communication, planning and direct interventions. She has lectured on the topic of Wildfire Smoke-Ready Communities at multiple conferences, workshops, and webinars in the United States and Canada, including a 2018 International Association of Wildland Fire conference, a 2018 Northwest Center for Public Health Practice Hot Topics in Practice webinar, the 2019 Health Effects Institute Annual Conference, a 2019 American Thoracic Society workshop, a 2019 Air & Waste Management Association conference, and a 2019 British Columbia Lung Association conference. Ms. Coefield has been a part of the Air Quality Program at the health department since 2010. She is the lead on smoke management, wildfire smoke response, and oxygenated fuels and also works on large projects, such as the PM_{10} Redesignation Request and the Carbon Monoxide Limited Maintenance Plan.

Delphine Farmer, PhD, is an associate professor of atmospheric chemistry at Colorado State University. Her research focuses on outdoor atmospheric and indoor chemistry with an emphasis on understanding the sources and sinks of reactive trace gases and particles and their effects on climate, ecosystems, and human health. Her recent work has focused on air chemistry in residential environments. She is a member of the NASEM Committee on Emerging Science on Indoor Chemistry. She received the Arnold and Mabel Beckman Young Investigator Award. Dr. Farmer earned her MS and PhD in chemistry from the University of California, Berkeley.

Andrea Ferro, PhD, is a professor in the Department of Civil and Environmental Engineering at Clarkson University. Her technical expertise is focused on indoor air quality and human exposure to particulate pollutants. She has also worked in the private industry, engineering consulting, and nonprofit sectors. The overall goal of her work is to improve human health by improving air quality through source control, ventilation and purification strategies, education, and regulatory policy. Dr. Ferro teaches undergraduate and graduate courses in environmental engineering, sustainable development, air pollution, and human exposure analysis. She is the faculty advisor for the Clarkson Environmental Conservation Organization (ECO) and the Clarkson student chapter of the American Association for Aerosol Research. She earned an MS in civil engineering and a PhD in civil and environmental engineering from Stanford University.

Elliott Gall, PhD, is an assistant professor in the School of Mechanical & Materials Engineering at Portland State University. His research and teaching investigate phenomena in built environments that affect indoor and urban environmental quality. Dr. Gall's Thermal and Fluid Science Group seeks to develop new approaches that improve building sustainability through an understanding of the intersection of building energy use, indoor air quality, and occupant wellbeing. Dr. Gall earned an MS in environmental and water resources engineering and a PhD in civil engineering from the University of Texas at Austin.

Lindsay T. Graham, PhD, is a research specialist at the Center for the Built Environment (CBE) of the University of California, Berkeley. She is a psychometrician and personality and social psychologist who specializes in the assessment of individuals in their daily environments and person-environment fit. Additionally, her research explores the ways in which human behaviors and personality influence the indoor air quality of an environment, and the physical and mental health consequences of these building-occupant interactions. She is working on ways to enhance assessment of human-building interactions, specifically through CBE's Occupant Indoor Environmental Quality Survey, a web-based tool that quantifies how buildings are performing from the perspective of the occupants. Dr. Graham earned a PhD in personality and social psychology from the University of Texas at Austin.

Stephanie Holm, MD, MPH, is codirector of the Western States Pediatric Environmental Health Specialty Unit at the University of California, San Francisco. She is board certified in both pediatrics and occupational/environmental medicine. She was the PI on the AQUA Study, a dual-cohort study of asthmatic children with and without cigarette exposure; it measured particulate matter levels in children's home environments to correlate these with other features of the household and behaviors of its occupants. Dr. Holm earned an MD from the University of Pittsburgh and an MPH in epidemiology from the University of California, Berkeley, and she is pursuing a PhD in epidemiology while continuing her research activities.

Cesunica Ivey, PhD, is an assistant professor in chemical and environmental engineering at the University of California, Riverside, and principal investigator of the Air Quality Modeling and Exposure Laboratory. She is also a member of the Bridging Regional Ecology, Aerosolized Toxins, and Health Effects (BREATHE) Center, a multidisciplinary collaborative. Dr. Ivey's research interests include source apportionment of fine particulate matter, regional air quality modeling for health applications, global atmospheric modeling, and environmental justice. She earned a PhD in environmental engineering from the Georgia Institute of Technology.

Howard Kipen, MD, MPH, is a professor of occupational and environmental health in the School of Public Health and director of Clinical Research and Occupational Medicine at the Environmental and Occupational Health Sciences Institute at Rutgers University. His research focuses on clinical and epidemiological studies of the health effects of ambient air pollution. He is a governor's appointee to the Public Employees' Occupational Safety and Health Review Commission, New Jersey Department of Labor, and member of the Public Health Scientific Advisory Board, New Jersey Department of Environmental Protection. He has served on several committees of the National Academies of Sciences, Engineering, and Medicine, including the Standing Committee on Medical and Epidemiological Aspects of Air Pollution on US Government Employees and Their Families. He received his MD from the University of California, San Francisco, and holds an MPH from Columbia University.

Kirsten Koehler, PhD, is an associate professor of environmental health and engineering in the Bloomberg School of Public Health at Johns Hopkins University. Her research seeks to improve exposure assessment methods to inform occupational and public health policy using direct-reading instrumentation to improve spatiotemporal exposure assessment. She is also developing novel aerosol samplers to improve the measurement of exposure to aerosolized particles and their health effects. In addition, Dr. Koehler is the principal investigator on a study exploring indoor exposure to traffic-related air pollution. She earned her BS from UCLA and MS and PhD from Colorado State University, all in atmospheric science.

Dusan Licina, PhD, is an assistant professor of indoor environmental quality at the School for Architecture, Civil, and Environmental Engineering at the Swiss Federal Institute of Technology Lausanne, where he leads the Human-Oriented Built Environment Lab. Dr. Licina's research focuses on the intersections between people and the built environment in order to ensure high indoor environmental quality for building occupants with minimum energy input. He specializes in air quality engineering, focusing on understanding of concentrations, dynamics and fates of air pollutants in buildings, and development and application of methods to quantitatively describe relationships between air pollution sources and consequent human exposures. His research interests also encompass optimization of building ventilation systems to improve air quality and thermal comfort in an energy-efficient manner. Dr. Licina earned an MS in mechanical engineering from the University of Belgrade, Serbia, and completed a joint doctorate degree from the National University of Singapore and the Technical University of Denmark.

Linsey Marr, PhD, is the Charles P. Lunsford Professor of Civil and Environmental Engineering at Virginia Tech and the principal investigator at the Center for Applied Interdisciplinary Research in Air. Her research interests include characterizing the emissions, fate, and transport of air pollutants in order to provide the scientific basis for improving air quality and health. She also conducts research on airborne transmission of infectious diseases. Dr. Marr was affiliated with the advisory board of Phylagen until January 2021 and currently consults for Smiths Detection, CrossFit, Inc., and the MITRE Corporation. She is a member of the National Academies' Board on Environmental Science and Toxicology and recently served on the planning committee for Airborne Transmission of SARS-CoV-2: A Virtual Workshop of the Academies' Environmental Health Matters Initiative and on the committee responsible for *Grand Challenges and Opportunities in Environmental Engineering and Science for the 21st Century*. In 2013 she received a New Innovator Award from the director of the National Institutes of Health. Dr. Marr earned a PhD in environmental engineering from the University of California, Berkeley.

Meredith C. McCormack, MD, MHS, is an associate professor of Medicine at the Johns Hopkins University School of Medicine with a joint appointment in environmental health and engineering at the Johns Hopkins Bloomberg School of Public Health. She is also associate program director of the Johns Hopkins Pulmonary and Critical Care Fellowship program and active in mentoring fellows and junior faculty. She has clinical expertise in asthma, COPD, and general pulmonary and critical care medicine, as well as pulmonary physiology and pulmonary function testing. She is the medical director of the Johns Hopkins Pulmonary Function Laboratory and vice chair of the American Thoracic Society Committee for Proficiency Standards in Pulmonary Function Testing. Her research focuses on the effect of environmental influences on underlying obstructive lung disease—specifically air pollution, diet, and obesity influences on COPD and asthma. She has been funded by the NIEHS and the EPA to conduct environmental cohort studies to understand the effects of indoor and outdoor air pollution on children and adults with underlying respiratory disease. Her work is largely focused in Baltimore City but has included rural areas of the State of Washington, Appalachia, and the Caribbean. Dr. McCormack earned an MD from Jefferson Medical College of Thomas Jefferson University and an MHS from the Johns Hopkins Bloomberg School of Public Health.

William Nazaroff, PhD, is the Daniel Tellep Distinguished Professor Emeritus of Engineering in the Department of Civil and Environmental Engineering at the University of California, Berkeley. His research, focused on the physics and chemistry of air pollutants in proximity to people, especially in indoor environments, involved exposure science, stressing the development and application of methods to better understand mechanistically the relationship between emission sources and human exposure to pollutants. Prior to his retirement, Dr. Nazaroff was editor in chief of *Indoor Air*; president of the Academy of Fellows in the International Society of Indoor Air Quality and Climate; and president of the American Association for Aerosol Research. He is coauthor of *Environmental Engineering Science* and has served on the National Academies Committee on the Effect of Climate Change on Indoor Air Quality and Public Health. Dr. Nazaroff received his master's in electrical engineering and computer science from the University of California, Berkeley, and holds a PhD in environmental engineering sciences from California Institute of Technology.

Jeffrey Siegel, PhD – *Dr. Siegel's biographic sketch is listed with the planning committee members above.*

Brett C. Singer, PhD, is a staff scientist, leader of the Indoor Environment Group, and principal investigator in the Energy Technologies Area of Lawrence Berkeley National Laboratory. He has conceived, conducted, and led research projects related to air pollutant emissions and physical-chemical processes in both outdoor and indoor environments, aiming to understand real-world processes and systems that affect air pollutant exposures. A major focus of his work has been indoor environmental quality and risk reduction in high-performance homes, with the goal of accelerating adoption of indoor air quality, comfort, durability, and sustainability measures into new homes and retrofits of existing homes. Key focus areas of this work are low-energy systems for filtration, smart ventilation, and mitigation approaches to indoor pollutant sources, including cooking. Dr. Singer codeveloped the population impact assessment modeling framework (PIAMF). He earned a PhD in civil and environmental engineering from the University of California, Berkeley.

Brent Stephens, PhD, is a professor and chair of the Department of Civil, Architectural, and Environmental Engineering at Illinois Institute of Technology (IIT). He is an expert in the fate and transport of indoor pollutants, building energy and environmental measurements, HVAC filtration, human exposure assessment, building energy simulation, and energy-efficient building design. Dr. Stephens runs the Built Environment Research Group at IIT, in which undergraduate students, graduate students, and postdoctoral researchers conduct research on energy efficiency and indoor air quality in buildings. His recent research projects include improving and applying methods to measure the infiltration of outdoor particulate matter and reactive gases into homes; measuring gas and particle emissions from desktop three-dimensional printers and evaluating emission control devices; measuring the in situ particle removal efficiency of HVAC filters in real environments; developing inexpensive, open source devices based on the Arduino platform for measuring and recording long-term indoor environmental and building operational data; and characterizing the energy and air quality impacts of higher-efficiency HVAC filters in central residential air conditioning systems. Dr. Stephens earned an MSE in environmental and water resources engineering and a PhD in civil engineering from the University of Texas at Austin.

Marina Vance, PhD – *Dr. Vance's biographic sketch is listed with the planning committee members above.*

Michael Waring, PhD, is a professor and department head of civil, architectural, and environmental engineering at Drexel University. His research exists at the intersection of environmental and architectural engineering. It focuses on indoor air quality and exposure, indoor aerosol and chemical modeling, and sustainable buildings. He believes that making buildings function more effectively is imperative to solving many societal challenges. Dr. Waring has received the NSF CAREER Award as well as the New Investigator Award from the American Society for Heating, Refrigeration and Air-Conditioning Engineers. He earned a BA in English and economics, a BS in architectural engineering, an MS in environmental and water resources engineering, and a PhD in civil engineering, all from the University of Texas at Austin.